Schriftenreihe der Institute für Systemdynamik (IDS) und optische Systeme (ISO)

Die „Schriftenreihe der Institute für Systemdynamik (ISD) und optische Systeme (ISO)“ deckt ein breites Themenspektrum ab: von angewandter Informatik bis zu Ingenieurswissenschaften. Die Institute für Systemdynamik und optische Systeme bilden gemeinsam einen Forschungsschwerpunkt der Hochschule Konstanz. Die Forschungsprogramme der beiden Institute umfassen informations- und regelungstechnische Fragestellungen sowie kognitive und bildgebende Systeme. Das Bindeglied ist dabei der Systemgedanke mit systemtechnischer Herangehensweise und damit verbunden die Suche nach Methoden zur Lösung interdisziplinärer, komplexer Probleme. In der Schriftenreihe werden Forschungsergebnisse in Form von Dissertationen veröffentlicht.

The “Series of the institutes of System Dynamics (ISD) and Optical Systems (ISO)” covers a broad range of topics: from applied computer science to engineering. The institutes of System Dynamics and Optical Systems form a research focus of the HTWG Konstanz. The research programs of both institutes cover problems in information technology and control engineering as well as cognitive and imaging systems. The connective link is the system concept and the systems engineering approach, i.e. the search for methods and solutions of interdisciplinary, complex problems. The series publishes research results in the form of dissertations.

Weitere Bände in der Reihe http://www.springer.com/series/16265

Simon Grimm

Directivity Based Multichannel Audio Signal Processing For Microphones in Noisy Acoustic Environments

Simon Grimm
Stuttgart, Germany

Dissertation Universität Ulm, 13. Juni 2018, u.d.T.: Simon Grimm "Directivity Based Multichannel Audio Signal Processing For Microphones in Noisy Acoustic Environments."

ISSN 2661-8087 ISSN 2661-8095 (electronic)
Schriftenreihe der Institute für Systemdynamik (IDS) und optische Systeme (ISO)
ISBN 978-3-658-25151-2 ISBN 978-3-658-25152-9 (eBook)
https://doi.org/10.1007/978-3-658-25152-9

Library of Congress Control Number: 2018967682

Springer Vieweg

This Springer Vieweg imprint is published by the registered company Springer Fachmedien Wiesbaden GmbH part of Springer Nature
The registered company address is: Abraham-Lincoln-Str. 46, 65189 Wiesbaden, Germany

Acknowledgements

First and foremost, I would like to thank my doctoral supervisor Prof. Dr. Jürgen Freudenberger for offering me the opportunity to join his research group at the Institute of System Dynamics (ISD) at the University of Applied Sciences (HTWG) Konstanz, Germany. Due to his efforts it became possible for me to work on my Ph.D. for the last three and a half years. I would like to thank him especially for the time and effort he has taken for the long and fruitful discussions we had. During my time at the institute I appreciated the valuableness of these discussions. I am very grateful for his advice and support which helped me gain my research skills.

I would also like to thank all the members of my research group which I met during my time at the institute. It would be really unfair to mention only a few people since everyone really made my time there a pleasure. The conversations during the coffee breaks in the kitchen were always fun, helped me to stay motivated and offered me a fresh mindset.

I would like to thank my family for all their help and support during my study time and my research period. Further, my gratitude goes to everyone that supported me during my Ph.D. project, especially Sarah. For proofreading this thesis, I would like to thank Jutta Seifriz.

Contents

Abstract

In this work, new multi-microphone signal processing strategies are examined that aim to achieve noise reduction and dereverberation. Therefore, narrow-band signal enhancement approaches are combined with broad-band processing in terms of directivity based beamforming.

Previously introduced formulations of the multichannel Wiener filter rely on the second-order statistics of the speech and noise signals. In this thesis, it is examined how additional knowledge about the location of a speaker as well as the microphone arrangement can be used in order to accomplish further noise reduction as well as dereverberation. This is achieved by new directivity based references for the generalized multichannel Wiener filter. For spatially distributed microphone arrangements that exploit the diversity of the sound field new references based on delay-and-sum beamforming are investigated. These improve the noise reduction and dereverberation capabilities compared with the standard speech distortion weighted multichannel Wiener filter. For closely spaced microphones differential beamforming is used to create directivity based references for the multichannel Wiener filter, which are able to suppress noise from specific room directions.

Since closely spaced microphone arrangements are often used in hearing aids, it is examined if the proposed directivity based references can be used in binaural applications, where the preservation of the binaural cues is an important aim. It is shown that the directivity based reference choices are able to preserve the binaural cues of a speech source, while the signal-to-noise ratio can be improved compared with the standard binaural multichannel Wiener filter.

Differential beamforming is very sensitive to wind noise. It is examined, how noise reduction is possible for closely spaced microphone arrays in case wind noise occurs. Due to the highly instationary wind noise terms, the estimation of the noise power spectral densities is a challenging task. By exploiting the different signal properties of speech and wind noise, a noise reduction approach is derived based on the decomposition of the multichannel Wiener filter which successfully reduces wind noise.

1 Introduction

In our daily lives, speech communication devices are omnipresent and have gained much interest in recent years. Widely available consumer products like smartphones, laptops, tablets or the most recently introduced smart loudspeakers are equipped with acoustic sensors that allow to perform a wide range of tasks in terms of speech signal capturing. For example, these include conversations with far end speakers, teleconferencing applications with multiple participants or the use of voice recognition software for speech to text applications, voice control for navigation systems or the utilization of world wide web services.

But also hearing-aids or hands-free communications in a car environment depend on acoustic sensors. The development of cost-effective micro-electromechanical system (MEMS) air pressure sensors in recent years allows to expand the number of equipped microphones per device without increasing the manufacturing costs significantly. A multi-microphone setup can potentially improve the quality of a desired signal source compared with a single microphone. This is achieved by forming a microphone arrangement to exploit the spatial properties of the sound field. Source direction sensitive microphone arrays can be realized by a narrow distance microphone spacing, while further distances between the acoustic sensors allow to sample the spatial sound field to benefit from signal diversity. Both approaches offer interesting possibilities, while the particular configuration depends on the specific use case.

While smartphones, smart televisions or laptops are equipped with multiple acoustic sensors, also hearing aids benefit from wireless link connections of the microphones at both ears to achieve diversity based signal processing. For many scenarios, the location of the desired speech source can vary over time. Therefore, the positioning of multiple acoustic sensors on several locations allows to cover a wider area for sound capturing compared with a single microphone.

S. Grimm, *Directivity Based Multichannel Audio Signal Processing For Microphones in Noisy Acoustic Environments*, Schriftenreihe der Institute für Systemdynamik (IDS) und optische Systeme (ISO), https://doi.org/10.1007/978-3-658-25152-9_1

Directional sensitive microphone arrangements are able to augment signals from certain incident directions while attenuating signal from other locations, which promise great capabilities in spatial noise suppression. However, all of these sound capturing approaches require a suitable signal combining strategy, which often additionally needs to be adaptive due to variations of the desired sound source position.

If we consider a variety of application scenarios for microphones, several factors can disturb the desired signal. For example, in the case of teleconferencing, the signal quality can be decreased due to effects caused by room acoustic influences. For hands-free communications in a car, the signal quality can be degraded by background noise. Undesired wind noise signal artifacts can occur at the microphones during conversations, which are held in windy conditions. In case of people wearing hearing aids, interfering background speakers can cause annoyance while holding talks, which leads to hearing fatigue and concentration problems regarding the conversation content.

Under ideal circumstances, the desired speech signal is best captured in a "get it right at the source" approach, which aims to prevent the speech signal to be degraded in the first place. For example, this can be achieved by using high quality acoustic sensors that have a linear frequency response, good transient preservation, provide low intrinsic noise and are able to handle sound pressure levels within a high dynamic range. Also suitable positioning of the microphones regarding the sound source helps to augment the desired signal in relation to background noise. For wind noise artifacts, a mechanical mesh of fabric can be used to dissipate turbulent air flow, while room influences can be reduced by proper acoustic treatment. However, in many cases this is not possible due to various reasons, which include manufacturing costs or environments which do not provide ideal properties regarding speech communication. Also the location of the desired signal source is sometimes simply not known in advance.

As a consequence regarding these non-idealities, signal processing algorithms are investigated in this work, which aim to combine several acoustic sensors in a suitable manner. The processing aims to augment the desired speech signals, while reducing the influence of unwanted disturbances. The use of more than one acoustic sensor allows to take the spatial sound field into account. The provided signal processing algorithms can potentially be used in a wide range of applications, due to the ubiquitous communication devices of all kinds. The specific problem statement of this thesis regarding the microphone combining is outlined in the next section.

1.1 Problem Statement

In this thesis, multichannel signal processing algorithms for microphones arrangements in noisy acoustic environments are examined. To achieve noise reduction, dereverberation and reduced speech distortion, the extension of narrow-band signal processing algorithms by broad-band approaches in terms of directivity-based beamforming is investigated.

One commonly used technique is the speech distortion weighted multichannel Wiener filter (SDW-MWF) [1, 2, 3, 4, 5], which considers a trade-off between speech distortion and noise reduction as a broad-band optimization criterion. The reference for the speech distortion thereby is the speech component in one microphone. The generalized multichannel Wiener filter [6] allows to expand this algorithm with a reference that can be designed by a combination of the individual microphone channels to form the overall transfer function. Recent approaches [7, 6] consider the second order statistics of the input signals to create a reference, which results in an improved output signal regarding the signal-to-noise ratio and speech distortion compared with the SDW-MWF.

In this work, additionally to the second order statistics, knowledge about the microphone arrangement and the location of the desired signal source is taken into account for the reference design. This allows to create a class of directivity based references that rely on classical beamforming approaches found in the literature [8, 9]. Based on the microphone arrangement and the specific signal disruptions, environmental dependent reference designs for signal improvement are investigated regarding their capability to improve the signal-to-noise ratio as well as to reduce acoustical influences.

Furthermore, the influence of these reference designs for hearing-aids, which use multichannel noise reduction algorithms, is investigated. Therefore, the influence of the reference designs on the preservation of the binaural cues, which are crucial for spatial hearing, is examined.

Besides background noise and reverberation, wind-induced disruptions are a common problem in hearing aids as well as in other communication applications. The highly non-stationary wind noise properties make it a challenging task to estimate the noise terms required for noise reduction. It is examined if the signal properties of speech and wind noise between closely spaced microphones can be exploited for an optimal signal combining to achieve wind noise reduction.

1.2 Structure of the Thesis

This work is organized in the following way: In chapter 2, state-of-the-art narrow-band and broad-band signal combining concepts for multichannel microphone setups, which are used in the following chapters, are introduced.

In chapter 3, the generalized multichannel Wiener filter is introduced. The generalization of the speech distortion weighted Wiener filter allows to acquire an overall acoustic transfer function not only by the selection of a single microphone channel, but by combining the individual microphone channels. It is shown that the overall acoustic transfer function has no impact on the narrow-band signal-to-noise ratio, but impacts the broadband signal-to-noise ratio. This generalization is important for this thesis. It allows to combine the state-of-the-art directivity based broadband beamforming approaches, as described in chapter 2, with the multichannel Wiener filter. This chapter has been published in [10].

In chapter 4, the generalized multichannel Wiener filter of chapter 3 is used to design the overall transfer function of the multichannel Wiener filter. Therefore, directivity based references are derived that exploit the spatial sound field to improve the broad-band signal-to-noise ratio as well as to reduce reverberation. Dependent on the microphone arrangement, suitable reference designs are derived. Differential beamforming references are proposed for closely spaced microphones. For spatially distributed acoustic sensors, delay-and-sum based references are introduced. The proposed references are applied in the context of a monaural hearing aid, a car environment and a classroom scenario. The results of this chapter are partly presented in [11, 10] and [12].

In chapter 5, the differential beamforming references derived in chapter 4 are investigated regarding their capability to preserve the binaural cues if they are applied in the context of binaural hearing aids. The binaural cues are important for a human being to localize signal sources in the spatial sound field. Therefore the binaural multichannel Wiener filter is introduced together with its generalization, similar to the generalized multichannel Wiener filter in chapter 3. The influence of the differential beamforming references on the binaural cues of the sound field is examined.

Differential beamforming and therefore the differential beamforming references derived in chapter 4 are sensitive to wind noise. For this reason, a wind noise reduction approach for closely spaced microphones is derived, based on the decomposition of the multichannel Wiener filter into a beamformer and a single channel post filter in chapter 6. The estimation of the highly non-stationary wind noise power spectral densities is obtained by exploiting the different correlation properties of speech and wind noise for closely spaced microphones. The results of this chapter are partly presented in [13] and [14].

In chapter 7, it is examined how multichannel noise reduction algorithms, as derived in the previous chapters, can be verified using background noise simulation. Therefore, a multiple input - multiple output equalization approach that uses more loudspeakers than microphones is derived. This equalization approach aims to preserve the spatial properties between the microphones if the pre-equalized signals are played back through the loudspeakers. This chapter has been published in [15]. Finally, a conclusion of this thesis is given in chapter 8.

2 Noise Reduction using Multichannel Signal Processing Approaches

In the following chapter, basic noise reduction techniques for multichannel microphone setups are derived. Since in this work directivity based signal combining approaches are investigated, the concepts of delay-and-sum beamforming as well as differential beamforming are introduced [9, 16, 17, 18, 19]. These broad-band signal processing techniques allow to create a directional response in terms of an incident angle dependent sensitivity. They have in common that they rely only on the information about the time difference of arrival regarding the desired signal source.

Furthermore, the well-known minimum variance beamforming [20, 21, 22] as well as the speech distortion weighted multichannel Wiener filter [1, 2, 3, 4] are presented. Both noise reduction approaches are quite similar regarding the beamforming, but differ by an additional post filter in the Wiener filter. By introducing relative transfer function beamforming, it can be shown that the multichannel Wiener filter as well as the minimum variance beamformer rely only on the second order statistics of the signals to achieve optimal noise reduction.

2.1 Signal Model and Notation

In this section, the signal model and the corresponding notation is introduced. The speech signal is often affected by acoustic influences and background noise in a typical real world environment. Background noise can contain several sources. For example, if a car environment is examined, noise is caused by the engine as well as rolling noise of the tires. Furthermore, wind noise at the microphones can be caused by open windows, fans or open convertible hoods. Another example is a cocktail party scenario, where babble noise terms

S. Grimm, *Directivity Based Multichannel Audio Signal Processing For Microphones in Noisy Acoustic Environments*, Schriftenreihe der Institute für Systemdynamik (IDS) und optische Systeme (ISO), https://doi.org/10.1007/978-3-658-25152-9_2

can appear. Regarding the acoustic influences, the speech signal can be affected by early reflections caused by room boundaries, which cause constructive and destructive interferences. These result in coloration of the magnitude spectrum as well as a non-minimum phase transfer function. Besides early reflections, the accumulated late reflections which cause reverberation need to be considered, since they degrade the speech intelligibility further.

The acoustic system is considered as a linear and time-invariant, which therefore can be described by an impulse response from the mouth-reference-point (MRP) of a speaker to the dedicated microphone. This allows the microphone signal $y_i(k)$ to be expressed as the convolution of the clean speech signal $x(k)$ at the MRP with an acoustic impulse response $h_i(k)$ plus an additive noise term $n_i(k)$

$$y_i(k) = s_i(k) + n_i(k) \tag{2.1}$$

$$s_i(k) = x(k) * h_i(k)\,, \tag{2.2}$$

where $*$ denotes the convolution, $s_i(k)$ the reverberant speech signal, i is the index of the ith microphone signal and k denotes the discrete time index. The signal model is also depicted in Figure 2.1.

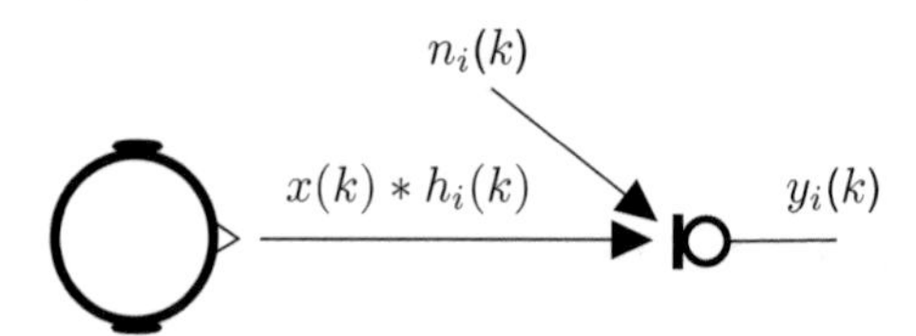

Figure 2.1: The signal model

To combine the signals for speech processing, they are filtered by a function $g_i(k)$ and then summed to form the output signal $z(k)$

$$z(k) = \sum_{i=1}^{M} y_i(k) * g_i(k)\,, \tag{2.3}$$

where M denotes the total number of microphones. Broad-band signal processing algorithms are commonly computed in the time domain. The filter functions $g_i(k)$ often consist of a scalar weighting constant as well as an optional delay, which is applied to the microphone signals. The time domain signal processing structure is depicted in Figure 2.2

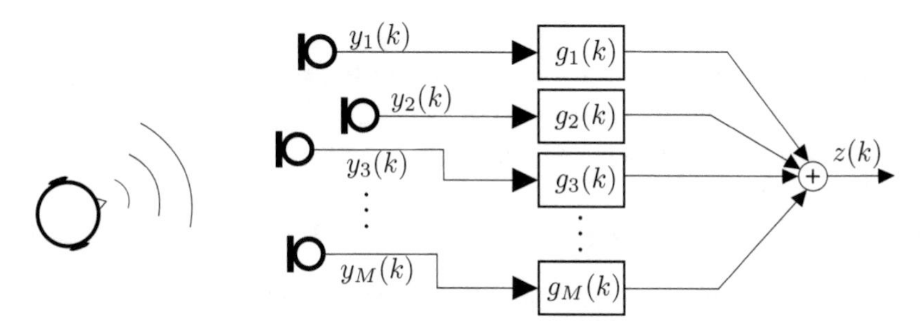

Figure 2.2: The system structure in the time domain with an arbitrary microphone arrangement

To formulate the narrow-band criteria signal processing algorithms, which are commonly described in the short time frequency domain, the microphone signals can be written as follows

$$Y_i(\eta,\nu) = S_i(\eta,\nu) + N_i(\eta,\nu) \tag{2.4}$$
$$S_i(\eta,\nu) = X(\eta,\nu)H_i(\nu)\,. \tag{2.5}$$

$Y_i(\eta,\nu)$, $X(\eta,\nu)$, and $N_i(\eta,\nu)$ correspond to the short time spectra of the time domain signals. $H_i(\nu)$ represents the acoutic transfer function (ATF) corresponding to the acoustic impulse response and $S_i(\eta,\nu)$ is the speech component at the ith microphone. η denotes the subsampled time index and ν the frequency bin index for a block length of L samples respectively. In the following these indices are often omitted for brevity. The short time spectra and the ATFs can be written as M-dimensional vectors:

$$\mathbf{S} = [S_1, S_2, \ldots, S_M]^T \tag{2.6}$$
$$\mathbf{N} = [N_1, N_1, \ldots, N_M]^T \tag{2.7}$$
$$\mathbf{H} = [H_1, H_2, \ldots, H_M]^T \tag{2.8}$$
$$\mathbf{Y} = [Y_1, Y_2, \ldots, Y_M]^T \tag{2.9}$$
$$\mathbf{Y} = \mathbf{S} + \mathbf{N} \tag{2.10}$$

^{T} denotes the transpose of a vector, * the complex conjugate and † the conjugate transpose. Vectors and matrices are written in bold and scalars are normal letters.

The speech and noise signals are assumed to be zero-mean random processes with the power spectral densities (PSDs) $\Phi^2_{S_i}$ and $\Phi^2_{N_i}$. Assuming a single speech source, the speech correlation matrix $\boldsymbol{R}_S$ has rank one and can be expressed as

$$\boldsymbol{R}_S = \mathbb{E}\left\{\mathbf{S}\mathbf{S}^\dagger\right\} = \Phi^2_X \mathbf{H}\mathbf{H}^\dagger , \tag{2.11}$$

where $\mathbb{E}\{\}$ denotes the mathematical expectation and Φ^2_X the PSD of the clean speech signal at the MRP. Similarly, $\boldsymbol{R}_N = \mathbb{E}\left\{\mathbf{N}\mathbf{N}^\dagger\right\}$ denotes the noise correlation matrix. It is further assumed that the speech and noise terms are orthogonal, so the input signal correlation matrix $\boldsymbol{R}_Y$ can be written as

$$\boldsymbol{R}_Y = \boldsymbol{R}_S + \boldsymbol{R}_N . \tag{2.12}$$

If we apply a signal processing algorithm to enhance the desired speech signal, this is obtained by filtering the microphone signals with a suitable filter function $\mathbf{G}$. Therefore, the output signal Z of an algorithm with the filter coefficients

$$\mathbf{G} = \left[G_1, G_2, \ldots, G_M\right]^T \tag{2.13}$$

is obtained by filtering and summing the microphone signals, i.e.,

$$\begin{aligned} Z &= \mathbf{G}^\dagger\mathbf{Y} = \mathbf{G}^\dagger\mathbf{S} + \mathbf{G}^\dagger\mathbf{N} \\ &= Z_S + Z_N \end{aligned} \tag{2.14}$$

where Z_S and Z_N denote the speech and the noise components at the output. The whole structure, from the mouth reference point of the speaker to the output of the signal processing algorithm, is depicted in Figure 2.3

2.2 Constant Directivity Beamforming

In the following section, microphone signal combining approaches, which are able to form a directional response, are introduced. Therefore an anechoic far

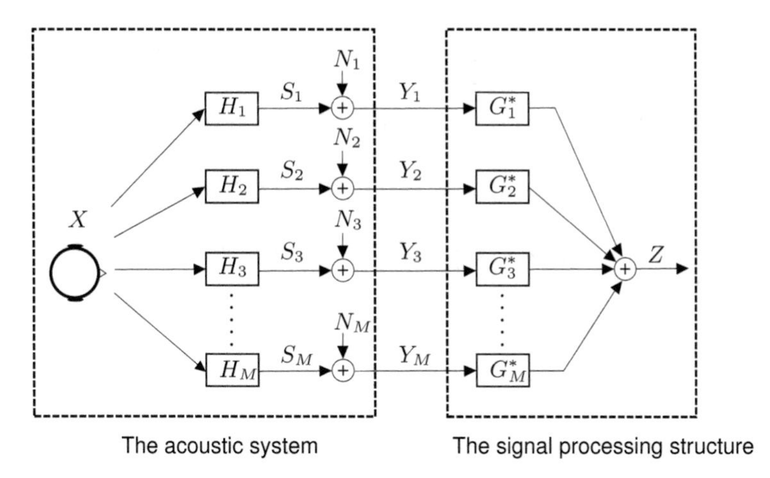

Figure 2.3: Complete system structure in the frequency domain

field environment is considered for the derivations. Two concepts that take the direction of arrival for the speech signal into account are presented, namely delay-and-sum beamforming as well as differential beamforming [9, 16, 17, 18, 19]. The former relies on the coherent combining of the desired signal, while the latter is able to create one or more spatial nulls for certain incidence angles. Under the assumption of a diffuse noise field, the concept of superdirective beamforming [23, 24] is presented, which aims to form a directional response to minimize the noise energy picked up by the sidelobes of the beam. Since the mentioned approaches are often associated with certain types of microphone array arrangements, two commonly used configurations, namely the broadside array and the endfire array, are presented in the following.

2.2.1 Microphone Arrangements

Microphones can be arranged in many ways to benefit from directivity based or spatial diversity combining [25]. For two-dimensional arrays, the acoustic sensors can be placed to form circular as well as linear arrays with equal or unequal spacing of the sensors. Also spherical array arrangements are possible, however, for constant directivity beamforming the scope is on planar linear arrays with equal spacing of the acoustic sensors in this thesis.

For planar arrays, only an azimuth angle θ is considered to describe the incident direction of a source signal. The angle dependent transfer characteristic $P_i(\nu, \theta)$ from a signal source to a single microphone can be described by

$$P_i(\nu, \theta) = p_i(\nu, \theta) e^{-j\phi_i(\nu, \theta)} , \tag{2.15}$$

where $p_i(\nu, \theta)$ and $\phi_i(\nu, \theta)$ denote the angle dependent magnitude and phase respectively.

Broadside Array

The broadside array consists of several microphones, which are arranged in line. The direction of arrival of the desired speech source ideally is perpendicular to the axis of the broadside array. The corresponding microphone arrangement is depicted in Figure 2.4.

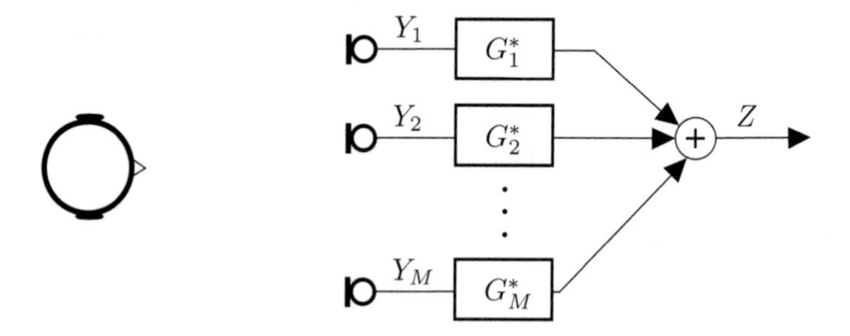

Figure 2.4: The broadside array

Endfire Array

Also the endfire array is an arrangement with the microphones positioned in line. However, the direction of arrival for the desired speech source is on axis with the microphones. Figure 2.5 depicts this configuration.

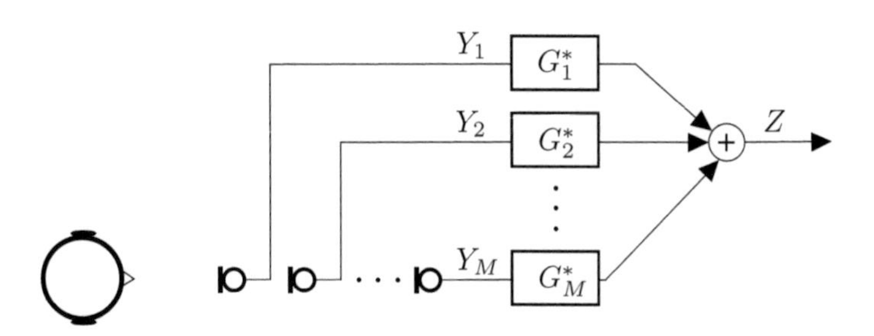

Figure 2.5: The endfire array

2.2.2 Delay-and-Sum Beamforming

Delay-and-sum beamforming is a combining approach that relies on coherent summing of the direct path of the desired speech source signal. This is achieved by delaying and weighting the microphone signals in a suitable manner before they are added. To implement this as a filter function, a steering vector is used to achieve a certain look direction of the beamformer, which depends on the incident angle θ of the source signal. The steering vector can be written as

$$\mathbf{G}^{\mathbf{DS}} = \frac{1}{M}[1, e^{j2\pi\nu\tau_2}, e^{j2\pi\nu\tau_3}, \ldots, e^{j2\pi\nu\tau_M}]^T , \tag{2.16}$$

where τ_i denotes the delay corresponding to the ith microphone. It can be calculated as

$$\tau_i = (i-1)\frac{cos(\theta)\delta}{c} , \tag{2.17}$$

where δ is the distance between two of the microphones in an equally spaced linear array and c is the speed of sound. Figure 2.6 depicts the setup for an angle dependent steering vector. Delay-and-sum beamforming can be used for broadside as well as for endfire microphone array configurations. Due to the coherent summing of the direct signal, incoherent reflections and noise terms are reduced in echoic environments, which makes the delay-and-sum approach suitable for dereverberation as well as noise reduction.

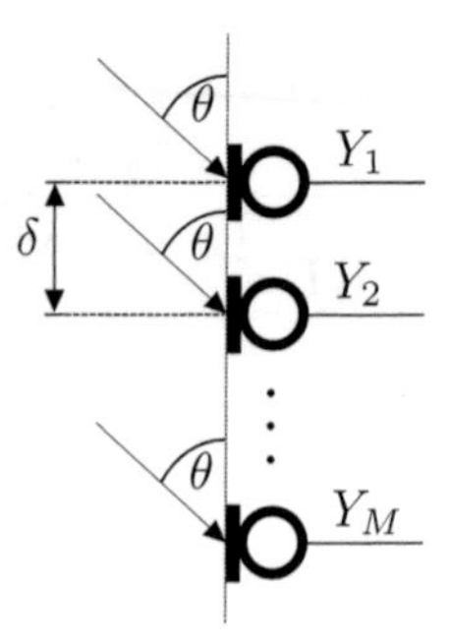

Figure 2.6: The microphone array - steering vector

2.2.3 Differential Beamforming

In contrast to delay-and-sum beamforming, which focuses on the look direction for a desired signal, differential beamforming allows to steer one or more nulls to suppress signals from a certain incident angle. The derivations for the differential microphone array are based on the work in [9]. Since differential arrays of higher order are quite challenging to realize due to non-idealities of the sensors, the focus is on first order microphone arrays. Anechoic farfield conditions are assumed, which means the acoustic transfer function vector $\mathbf{H}$ only consists of the direct paths from the mouth reference point of a speaker to the microphones. This allows to define the elements of the steering vector $\mathbf{d}(\nu, \theta)$ using (2.15) and (2.16)

$$d_i(\nu, \theta) = P_i(\nu, \theta) e^{-j2\pi\nu\tau_i}, \tag{2.18}$$

which is required to formulate the constraints for the directional response. One constraint is the distortionless response for the direction of the desired signal ($\theta_d = 0°$), while the other constraint is to create a spatial null in a chosen direction $\theta_n \in [90°, 180°]$.

$$\mathbf{G}^{\mathbf{DIFF}}(\nu)^{\dagger}\mathbf{d}(\nu, 0°) = 1 \tag{2.19}$$

$$\mathbf{G}^{\mathbf{DIFF}}(\nu)^{\dagger}\mathbf{d}(\nu, \theta_n) = 0. \tag{2.20}$$

For a first order differential array, consisting of $M = 2$ microphones in endfire configuration, the solution is given by (for more details, see [9])

$$\mathbf{G}^{\mathbf{DIFF}} = \frac{C}{\nu}[1, -e^{j2\pi\nu\tau}]^T . \quad (2.21)$$

C denotes a constant makeup gain factor, which depends on the distance between the microphones. The delay τ can vary within $0 \leq \tau \leq \tau_0$. The time delay τ_0 is defined as the propagation time from one microphone to the other

$$\tau_0 = \frac{\delta}{c} . \quad (2.22)$$

Note that the factor $\frac{1}{\nu}$ in equation (2.21) is a lowpass filter to equalize the highpass characteristic of the differential microphone array output. Dependent on τ various beam patterns can be formed, which are described by the angle dependent magnitude transfer characteristic $P_{diff}(\theta)$ of the differential array output (for more details, see [25]):

$$|P_{diff}(\theta)| = cos\left(\frac{\theta}{180^\circ}\pi\right) + \frac{\tau}{\tau_0} . \quad (2.23)$$

Alternatively, the beam pattern can be described by

$$P_{diff}(\theta) = (1 - b) + b \cdot cos\left(\frac{\theta}{180^\circ}\pi\right) , \quad (2.24)$$

where b is a real value ($b \in [0, 1]$), which can be interpreted as a weighting between an omnidirectional ($b = 0$) and a dipole ($b = 1$) transfer characteristic. Table 2.1 shows the values of $\frac{\tau}{\tau_0}$ and b for the corresponding beam patterns.

Table 2.1: Selected directional beam patterns for $\frac{\tau}{\tau_0}$ and b

Beam pattern	$\frac{\tau}{\tau_0}$	b
Omnidirectional	-	0
Dipole	0	1
Cardioid	1	0.5
Hypercardioid	$\frac{1}{3}$	0.75
Supercardioid	$\frac{2}{3}$	0.6

The resulting polar plots are depicted in Figure 2.7.

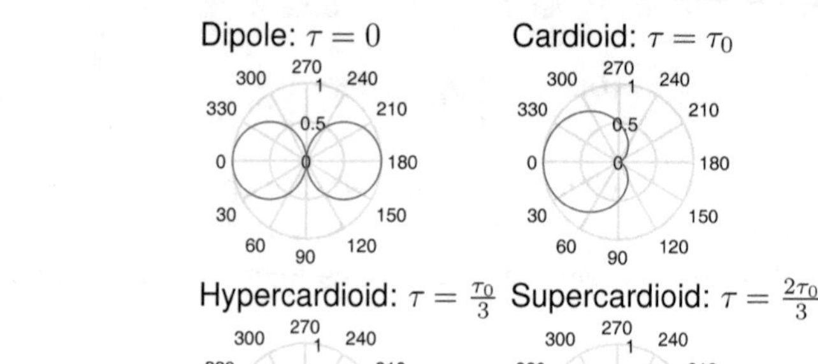

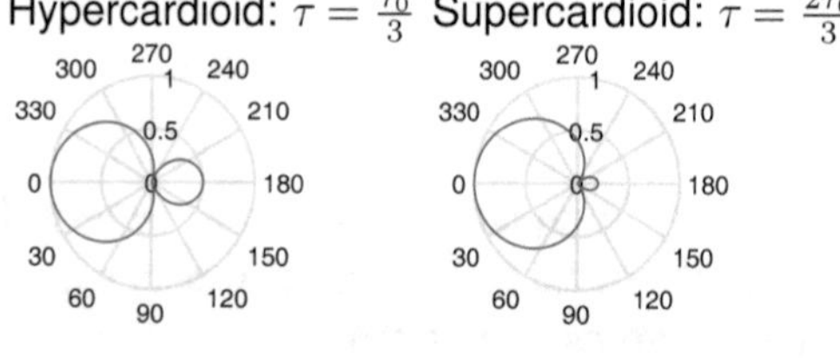

Figure 2.7: Differential array beam patterns for varying values of τ

As can be observed, the nulls can be steered from incident angles of 90° (270°) to 180° and all values in between, which allows a blocking of signals coming from these directions. Due to the restrictions of the steerable null, endfire arrays are used for differential beamforming, where the direction of the desired speech source ($\theta = 0^\circ$) is assumed to be known.

The performance measures regarding signal-to-noise ratio improvement as well as the ability to suppress uniformly distributed noise are defined by the array gain and the directivity factor [23].

Array Gain

The array gain describes the signal-to-noise ratio improvement at the output of the array compared to the input signal at one reference microphone denoted by the index ref. It is defined as

$$AG = \frac{\gamma^{out}}{\gamma^{in}_{ref}}, \tag{2.25}$$

where the input SNR at the reference microphone is defined as

$$\gamma_{ref}^{in} = \frac{\Phi_{S_{ref}}^2}{\Phi_{N_{ref}}^2} \tag{2.26}$$

and the SNR at the beamformer output as

$$\gamma^{out} = \frac{\mathbf{G}^{\dagger} \boldsymbol{R}_S \mathbf{G}}{\mathbf{G}^{\dagger} \boldsymbol{R}_N \mathbf{G}} . \tag{2.27}$$

The array gain can also be described by the filter function vector as

$$AG = \frac{|\mathbf{G}^{\dagger} \mathbf{d}|^2}{\mathbf{G}^{\dagger} \tilde{\boldsymbol{R}}_N \mathbf{G}} . \tag{2.28}$$

$\tilde{\boldsymbol{R}}_N$ is the normalized noise correlation matrix ($\tilde{\boldsymbol{R}}_N = \boldsymbol{R}_N / \Phi_{N_{ref}}^2$), similar as described in [24].

Directivity Factor

The directivity factor is a performance measure, which indicates the signal-to-noise ratio improvement for a directional microphone compared to an omni-directional microphone in a diffuse noise field described by the noise correlation matrix $\boldsymbol{R}_N^{diff}$. Quite similar to (2.28), it is defined as

$$DF = \frac{|\mathbf{G}^{\dagger} \mathbf{d}|^2}{\mathbf{G}^{\dagger} \boldsymbol{R}_N^{diff} \mathbf{G}} . \tag{2.29}$$

For an array consisting of two microphones, the hypercardiod beam pattern theoretically offers the best directivity factor (4.8 dB), closely followed by the supercardiod pattern (4.6 dB) [25].

2.2.4 Super Directive Beamforming

Often the problem occurs that more interfering sources than steerable nulls of a differential microphone array are present. The problem gets even worse in environments with strong reverberation. Instead of steering several nulls, the superdirective beamformer [23, 24] aims to minimize the sidelobes and narrows the mainlobe of the beam pattern to obtain an optimal directivity factor. If a steering vector $\mathbf{d}(\nu, \theta)$, as introduced in (2.18), is used, it allows to steer the beam in the look-direction of the desired speech source by applying the appropriate delays as described in (2.17).

Further, the noise field is assumed to be diffuse ($\boldsymbol{R}_N = \boldsymbol{R}_N^{diff}$). Under the constraint of a distortionless response in the direction of the speech source ($\mathbf{G^{SD}}^\dagger \mathbf{d}(\nu, 0°) = 1$), the minimization problem can be described as

$$\mathbf{G} = \underset{\mathbf{G}}{\operatorname{argmin}} \; \mathbf{G}^\dagger \boldsymbol{R}_N^{diff} \mathbf{G} \tag{2.30}$$

$$\text{subject to: } \mathbf{G}^\dagger \mathbf{d} = 1 \tag{2.31}$$

and the solution of the superdirective beamformer is given by (for more details, see [24])

$$\mathbf{G^{SD}} = \frac{\left(\boldsymbol{R}_N^{diff}\right)^{-1} \mathbf{d}}{\mathbf{d}^\dagger \left(\boldsymbol{R}_N^{diff}\right)^{-1} \mathbf{d}} \,. \tag{2.32}$$

Note that for a microphone array consisting of $M = 2$ microphones the superdirective beam pattern is equal to the hypercardioid beamformer as depicted in Figure 2.7 since both achieve the optimal directivity factor for this number of microphones in a diffuse noise field.

2.3 Minimum Variance Beamforming

In the following, narrow-band signal combining approaches for multiple microphone signals with respect to noise reduction and dereverberation capabilities are discussed. Therefore linear constrained minimum variance (LCMV) beamforming is derived, followed by the derivation of the minimum variance distortionless response (MVDR) beamformer. The MVDR beamformer is capable of reducing noise while also perfectly equalizing the room acoustic influences if $\boldsymbol{R}_N$ can be estimated sufficiently well and the acoustic transfer function vector $\mathbf{H}$ is known exactly. The derivation of the MVDR beamformer is based on the work in [20, 21, 22], where it was first introduced by [20]. In [21, 22] the MVDR beamformer is applied in the context of room acoustics. It can be shown that it is a special case of the linear constrained minimum variance (LCMV) beamformer, which is proposed in [26]. By setting the linear constraint equal to one of the acoustic transfer functions of the microphone signals, still maximum signal-to-noise ratio beamforming is achieved, while only knowledge of the signal correlation matrices $\boldsymbol{R}_S$ and $\boldsymbol{R}_N$ is required. However, the capability to equalize the room acoustic influences has to be sacrificed.

2.3.1 The Linear Constrained Minimum Variance Beamformer

Under the assumption that the speech and noise signals are uncorrelated, the error signal between the output of a beamformer and a speech signal which is parametrized by an arbitrary transfer function $\tilde{H}_d$ can be written as

$$\begin{aligned} \epsilon &= \mathbf{G^{LCMV}}^\dagger \mathbf{Y} - \tilde{H}_d X && (2.33) \\ &= (\mathbf{G^{LCMV}}^\dagger \mathbf{H} - \tilde{H}_d) X + \mathbf{G^{LCMV}}^\dagger \mathbf{N}\,, && (2.34) \end{aligned}$$

which can be placed in the mean squared error function given by

$$\begin{aligned} \xi(\mathbf{G^{LCMV}}) &= \mathbb{E}\left\{|\epsilon|^2\right\} && (2.35) \\ &= |(\mathbf{G^{LCMV}}^\dagger \mathbf{H} - \tilde{H}_d)|^2 \Phi_X^2 && (2.36) \\ &\quad + \mathbf{G^{LCMV}}^\dagger \boldsymbol{R}_N \mathbf{G^{LCMV}}\,. \end{aligned}$$

The minimization of the cost function in the minimum mean squared error sense can be written as

$$\begin{aligned} \mathbf{G^{LCMV}} &= \underset{\mathbf{G^{LCMV}}}{\operatorname{argmin}}\, \mathbf{G^{LCMV}}^\dagger \boldsymbol{R}_N \mathbf{G^{LCMV}} && (2.37) \\ \text{subject to: } \mathbf{G^{LCMV}}^\dagger \mathbf{H} &= \tilde{H}_d && (2.38) \end{aligned}$$

where $\tilde{H}_d$ introduces a degree of freedom in terms of a linear constraint to the optimization problem.

The solution to this optimization problem, including the linear constraint, is given by the LCMV beamformer

$$\mathbf{G}^{\mathbf{LCMV}} = \frac{\boldsymbol{R}_N^{-1}\mathbf{H}}{\mathbf{H}^{\dagger}\boldsymbol{R}_N^{-1}\mathbf{H}}\tilde{H}_d^* . \tag{2.39}$$

2.3.2 The Minimum Variance Distortionless Response Beamformer

By setting $\tilde{H}_d = 1$, a distortionless response for the speech signal in the desired look direction is achieved by

$$\mathbf{G}^{\mathbf{MVDR}} = \frac{\boldsymbol{R}_N^{-1}\mathbf{H}}{\mathbf{H}^{\dagger}\boldsymbol{R}_N^{-1}\mathbf{H}} , \tag{2.40}$$

which is the widely known MVDR beamformer. However, $\mathbf{G}^{\mathbf{MVDR}}$ requires knowledge of the acoustic transfer functions, which is mostly not available in practice, since the blind estimation of acoustic transfer functions in noisy conditions is a challenging task and so far an unsolved problem.

2.3.3 The Relative Transfer Function Minimum Variance Beamformer

By setting $\tilde{H}_d$ to an explicit reference channel ($\tilde{H}_d \in \{H_1, \dots, H_M\}$), the knowledge of the correlation matrices $\boldsymbol{R}_S$ and $\boldsymbol{R}_N$ is sufficient to obtain an optimum beamformer regarding the narrow-band signal-to-noise ratio as can be shown in the following. However, the acoustic transfer functions cannot be equalized with this approach.

If $\tilde{H}_d$ is set to an explicit reference channel ($\tilde{H}_d = H_{ref}$), then the beamformer can be written as

$$\mathbf{G}^{\mathbf{RTF-MV}} = \frac{\boldsymbol{R}_N^{-1}\mathbf{H}}{\mathbf{H}^{\dagger}\boldsymbol{R}_N^{-1}\mathbf{H}} H_{ref}^* \tag{2.41}$$

$$= \frac{\boldsymbol{R}_N^{-1}\mathbf{H}}{\mathbf{H}^{\dagger}\boldsymbol{R}_N^{-1}\mathbf{H}} \mathbf{H}^{\dagger}\mathbf{u} , \tag{2.42}$$

where $\mathbf{u}$ is a vector of length M which consists of zeros and a single one at the corresponding position of the selected reference channel.

By expanding the $\mathbf{G}^{\mathbf{RTF-MV}}$ beamformer in the nominator and the denominator with the clean speech PSD Φ_X^2, we obtain

$$\mathbf{G}^{\mathbf{RTF-MV}} = \frac{\Phi_X^2 \boldsymbol{R}_N^{-1}\mathbf{H}}{\Phi_X^2 \mathbf{H}^\dagger \boldsymbol{R}_N^{-1}\mathbf{H}} \mathbf{H}^\dagger \mathbf{u} \tag{2.43}$$

$$= \frac{\Phi_X^2 \boldsymbol{R}_N^{-1}\mathbf{H}\mathbf{H}^\dagger}{\mathrm{tr}\left(\Phi_X^2 \mathbf{H}^\dagger \boldsymbol{R}_N^{-1}\mathbf{H}\right)} \mathbf{u} \tag{2.44}$$

$$= \frac{\boldsymbol{R}_N^{-1}\boldsymbol{R}_S}{\mathrm{tr}\left(\boldsymbol{R}_N^{-1}\boldsymbol{R}_S\right)} \mathbf{u}\,, \tag{2.45}$$

where $\mathrm{tr}(\cdot)$ denotes the trace operator. It should be noted that $\mathbf{G}^{\mathbf{RTF-MV}}$ now only depends on knowledge of the speech and noise signal correlation matrices $\boldsymbol{R}_S$ and $\boldsymbol{R}_N$ and no further knowledge is required for the signal combining.

2.4 The Speech Distortion Weighted Multichannel Wiener Filter

In this section, the speech distortion weighted multichannel Wiener filter (SDW-MWF) is derived based on the minimum mean squared error (MMSE) criterion, similar to the LCMV beamformer in section 2.3. This derivation is based on the work of [1, 2, 3, 4]. In [1], the speech distortion weighted multichannel Wiener was proposed including a trade-off parameter which allows to control the amount of noise reduction in relation to linear speech signal distortion. Therefore speech distortion is explicitly taken into account in the optimization process.

2.4.1 Minimum Mean Squared Error Solution

Again the MMSE criterion is taken to minimize the mean squared error between a reference signal and the output of the signal processing algorithm. This MMSE solution results in the speech distortion weighted multichannel Wiener filter. The resulting cost function can be written as

$$\mathbf{G}^{\mathbf{MWF}} = \underset{\mathbf{G}^{\mathbf{MWF}}}{\mathrm{argmin}}\, \mathbb{E}\left\{ |\mathbf{G}^{\mathbf{MWF}\dagger}\mathbf{Y} - X\tilde{H}_d|^2 \right\}. \tag{2.46}$$

The property that noise and speech signals are uncorrelated can be exploited to write the MMSE signal energy $\mathbb{E}\left\{|\epsilon|^2\right\}$ as

$$\begin{aligned}\mathbb{E}\left\{|\epsilon|^2\right\} &= \mathbb{E}\left\{|\epsilon_x|^2\right\} + \mathbb{E}\left\{|\epsilon_n|^2\right\} && (2.47)\\ &= \mathbb{E}\left\{|\mathbf{G}^{\mathbf{MWF}\dagger}\mathbf{S} - X\tilde{H}_d|^2\right\} + \mathbb{E}\left\{|\mathbf{G}^{\mathbf{MWF}\dagger}\mathbf{N}|^2\right\}. && (2.48)\end{aligned}$$

The separation of the speech and noise signal energy can be utilized to include the trade-off parameter μ in the minimization process which was introduced in [1]. This leads to the minimization formula

$$\mathbf{G}^{\mathbf{MWF}} = \underset{\mathbf{G}^{\mathbf{MWF}}}{\operatorname{argmin}}\, \mathbb{E}\left\{|\mathbf{G}^{\mathbf{MWF}\dagger}\mathbf{S} - X\tilde{H}_d|^2\right\} + \mu\mathbb{E}\left\{|\mathbf{G}^{\mathbf{MWF}\dagger}\mathbf{N}\right\}, \quad (2.49)$$

where the solution is the speech distortion weighted multichannel Wiener filter

$$\mathbf{G}^{\mathbf{MWF}} = (\boldsymbol{R}_S + \mu\boldsymbol{R}_N)^{-1}\Phi_X^2\mathbf{H}\tilde{H}_d^* \quad (2.50)$$

which allows a trade-off between speech distortion and noise suppression by the parameter μ. Further, $\tilde{H}_d$ can be set to an arbitrary reference channel if we use a vector $\mathbf{u}$, similar to (2.43) in section 2.3.3, to select the corresponding transfer function in $\mathbf{H}$.

$$\begin{aligned}\mathbf{G}^{\mathbf{MWF}} &= (\boldsymbol{R}_S + \mu\boldsymbol{R}_N)^{-1}\Phi_X^2\mathbf{H}\mathbf{H}^\dagger\mathbf{u} && (2.51)\\ &= (\boldsymbol{R}_S + \mu\boldsymbol{R}_N)^{-1}\boldsymbol{R}_S\mathbf{u} && (2.52)\end{aligned}$$

Similar to (2.43), $\mathbf{G}^{\mathbf{MWF}}$ depends only on the speech and noise correlation matrices $\boldsymbol{R}_S$ and $\boldsymbol{R}_N$, so no further knowledge is necessary for this multichannel noise reduction approach.

2.4.2 Decomposition of the Multichannel Wiener Filter

In [27], [7] and many others it is shown that the multichannel Wiener filter can be decomposed in a MVDR beamformer, a single channel Wiener post filter and the overall transfer function $\tilde{H}_d$ using the matrix inversion lemma

$$\mathbf{G}^{\mathbf{MWF}} = \frac{\Phi_X^2}{\Phi_X^2 + \mu(\mathbf{H}^\dagger \boldsymbol{R}_N^{-1}\mathbf{H})^{-1}} \frac{\boldsymbol{R}_N^{-1}\mathbf{H}}{\mathbf{H}^\dagger \boldsymbol{R}_N^{-1}\mathbf{H}} \tilde{H}_d^* \tag{2.53}$$

$$= G^{WF}\, \mathbf{G}^{\mathbf{MVDR}}\, \tilde{H}_d^* . \tag{2.54}$$

This means, the spatial filtering is performed by the MVDR beamformer whereas additional broad-band noise reduction is achieved by the spectral filtering of the Wiener post filter. The overall transfer function is determined by $\tilde{H}_d$. Therefore the LCMV beamformer, as derived in equation (2.39), and the MWF only differ by the single channel Wiener post filter which is defined as

$$G^{WF} = \frac{\Phi_X^2}{\Phi_X^2 + \mu\Phi_{N,MVDR}^2} \tag{2.55}$$

where $\Phi_{N,MVDR}^2$ is the noise PSD at the output of $\mathbf{G}^{\mathbf{MVDR}}$

$$\Phi_{N,MVDR}^2 = \mathbf{G}^{\mathbf{MVDR}\dagger} \boldsymbol{R}_N \mathbf{G}^{\mathbf{MVDR}} \tag{2.56}$$

$$= (\mathbf{H}^\dagger \boldsymbol{R}_N^{-1}\mathbf{H})^{-1} . \tag{2.57}$$

2.5 Summary

In this chapter, state-of-the-art signal processing methods for multichannel noise reduction were derived. Required to describe the signal processing problem, the signal model and its corresponding notation in the time and frequency domain were introduced. Two commonly used planar and linear microphone arrangements of equally spaced sensors were presented, namely the broadside array and the endfire array. By introducing a class of constant directivity beamforming approaches, i.e. the delay-and-sum beamformer as well as differential microphone array processing, broad-band noise reduction algorithms were derived that are able to form a directional response for the beamformer output. These methods only require knowledge of the direction of arrival for the desired speech source. By additionally taking information about the correlation properties of the noise field into account, the superdirective beamformer was introduced which aims to minimize the beam pattern sidelobe energy for a diffuse noise field.

Besides the constant directivity beamforming approaches, narrow-band signal processing algorithms were introduced which are based on the solution of a minimum mean squared error optimization problem. One of those methods is the well-known minimum variance beamforming. The derivation of the MVDR beamformer shows that this approach is able to achieve optimal narrow-band beamforming while perfectly equalizing the acoustic system. However, this requires knowledge of the acoustic transfer functions, which is hardly available in practice. With the relative transfer function minimum variance beamformer, it can be shown that optimal narrow-band beamforming is still possible without knowledge of the acoustic transfer functions, however, the capability to equalize the acoustic system has to be sacrificed. Despite the lost capability, this approach has the advantage that only knowledge of the correlation properties of the speech and noise signals is required.

Similar, the SDW-MWF was derived based on the minimum mean squared error criterion. It was shown that the property of uncorrelated speech and noise signals can be exploited to include an additional parameter in the minimization process, which allows to set a trade-off between speech distortion and noise reduction.

Based on the matrix inversion lemma, it was presented that the multichannel Wiener filter can be decomposed into an MVDR beamformer, a single channel post filter and a resulting overall transfer function. By comparing the SDW-MWF with the LCMV beamformer, they only differ by the single channel Wiener post filter, which has no influence on the narrow-band SNR but has the capability to gain additional broad-band output SNR improvement.

3 The Generalized Multichannel Wiener Filter

In the past few years, research on speech enhancement using acoustic sensor networks consisting of spatially distributed microphones has gained significant interest [28, 29, 30, 31, 7, 32, 6, 33, 34, 35, 36, 37]. Compared with a microphone array at a single position, spatially distributed microphones are able to acquire more information about the sound field. The usage of spatially distributed microphones allows to employ combining techniques for speech quality improvement in reverberant and noisy conditions. Several methods were introduced that use an explicit reference channel. These include the relative transfer function - generalized sidelobe canceler (RTF-GSC) [38], the MVDR beamformer [21] and the speech distortion weighted - multichannel Wiener filter [1, 2, 3, 4, 5].

The MWF, which is a well-established technique for speech enhancement, produces a minimum-mean-squared error estimate of an unknown desired signal as derived in chapter 2. The desired signal of the standard MWF (S-MWF) is usually the speech component in one of the microphone signals, referred to as the reference microphone signal. However, for spatially distributed microphones the selection of the reference microphone may have a large influence on the performance of the MWF since it depends on the positions of the speech and noise sources as well as the microphones [39, 7, 6, 32].

With the S-MWF, the overall transfer function from the speaker to the output of the MWF equals the acoustic transfer function from the speaker to the reference microphone. Hence, the reference microphone selection determines the amount of speech distortion. Moreover, the overall transfer function has an impact on the broadband output SNR of the MWF [39]. This raises the need to select a suitable reference. Therefore, the generalized MWF (G-MWF) was proposed in order to improve the broadband output SNR [6] (see also [32]). With the G-MWF, the speech reference is not chosen by selecting one of the microphone channels as in the standard multichannel Wiener filter, but by a

S. Grimm, *Directivity Based Multichannel Audio Signal Processing For Microphones in Noisy Acoustic Environments*, Schriftenreihe der Institute für Systemdynamik (IDS) und optische Systeme (ISO), https://doi.org/10.1007/978-3-658-25152-9_3

combining of the channels to create a new overall transfer function in order to improve the broad-band output SNR.

In this chapter, the G-MWF formulation is presented in section 3.1. It is shown that the G-MWF includes the S-MWF as a special case. The G-MWF derivation is followed by a discussion of the narrow-band and broad-band output signal-to-noise ratio in section 3.2 to examine the difference and the dependence of the broad-band output SNR on the overall transfer function.

This chapter has been presented in [10].

3.1 The Generalized MWF

The SDW-MWF, as proposed in section 2.4, aims to estimate the speech signal of a chosen microphone channel. It is commonly implemented as

$$\mathbf{G}^{\mathbf{MWF}} = (\boldsymbol{R}_S + \mu \boldsymbol{R}_N)^{-1} \boldsymbol{R}_S \mathbf{u} \,, \tag{3.1}$$

where $\mathbf{u}$ is a vector that selects the reference microphone, i.e. the vector $\mathbf{u}$ contains a single one and all other elements are zero

$$\mathbf{u} = [0, \ldots, 1, \ldots, 0]^T \,. \tag{3.2}$$

This MWF realization is referred to as the S-MWF in the following. Therefore for the S-MWF, the overall transfer function is equal to the ATF of a reference microphone, i.e., $\tilde{H}_d = H_{ref}$. Since $\boldsymbol{R}_S$ is a rank one matrix, it should be noted that any non-zero vector $\mathbf{u}$ achieves the same (optimal) narrow-band output SNR and therefore it is independent of the choice of $\mathbf{u}$. This implies that the vector $\mathbf{u}$ can be chosen in an arbitrary way to influence the overall transfer function $\tilde{H}_d$.

In [6], the generalized MWF was presented, which allows to define a speech reference for the MWF by the elements u_i of the vector $\mathbf{u}$. This is achieved by forming a weighted sum of the speech components in the different microphones with the phase of the speech component in the reference microphone signal.

The vector $\mathbf{u}$ can be used to define the desired overall response as

$$\tilde{H}_d = \mathbf{u}^\dagger \mathbf{H} = \sum_i {u_i}^* \cdot H_i \text{ for } u_i \in \mathbb{C}\,. \tag{3.3}$$

By using the decomposition of the MWF, it can be seen that $\tilde{H}_d$ is indeed the resulting overall transfer function

$$\mathbf{G} = \frac{\Phi_X^2}{\Phi_X^2 + \mu(\mathbf{H}^\dagger \boldsymbol{R}_N^{-1}\mathbf{H})^{-1}} \frac{\boldsymbol{R}_N^{-1}\mathbf{H}}{\mathbf{H}^\dagger \boldsymbol{R}_N^{-1}\mathbf{H}} \tilde{H}_d^* \tag{3.4}$$

$$= G^{WF}\,\mathbf{G}^{\mathbf{MVDR}}\,\tilde{H}_d^*\,, \tag{3.5}$$

since without noise reduction, i.e. for $\mu = 0$, the overall transfer function equals $\tilde{H}_d$ because $\mathbf{G}^{\mathbf{MVDR}}$ has a unity gain transfer function.

In this case, the speech output signal of the processing algorithm can be written as

$$Z_S = \tilde{H}_d \cdot X\,. \tag{3.6}$$

3.2 On the Output SNR

In the following, the signal-to-noise ratio for the input signal as well as the filtered output of the signal processing scheme is considered. Therefore it is distinguished between the narrow-band as well as the broad-band SNR for the multichannel Wiener filter. It can be shown that the selection of an arbitrary transfer function $\tilde{H}_d$ has no influence on the narrow-band SNR. However, the resulting output SNR is affected, as is derived in the following.

3.2.1 Narrow-band SNR Considerations

The narrow-band input SNR γ^{in} is defined as

$$\gamma^{in}(\nu) = \frac{\mathbb{E}\left\{\mathbf{S}\mathbf{S}^\dagger\right\}}{\mathbb{E}\left\{\mathbf{N}\mathbf{N}^\dagger\right\}} = \frac{\boldsymbol{R}_S}{\boldsymbol{R}_N} = \frac{\Phi_X^2 \mathbf{H}\mathbf{H}^\dagger}{\boldsymbol{R}_N}\,. \tag{3.7}$$

Therefore it follows for the narrow-band input SNR at the ith microphone as

$$\gamma_i^{in}(\nu) = \frac{\Phi_X^2 |H_i|^2}{\Phi_{N_i}^2} \tag{3.8}$$

which clearly indicates that γ_i^{in} depends on the PSD of the noise signal as well as on the acoustic transfer function at the ith microphone.

Now, the narrow-band SNR of the MWF output is considered, which is defined as

$$\gamma^{out}(\nu) = \frac{\mathbb{E}\left\{|Z_S(\nu)|^2\right\}}{\mathbb{E}\left\{|Z_N(\nu)|^2\right\}} = \frac{\mathbf{G^{MWF}}^\dagger \boldsymbol{R}_S \mathbf{G^{MWF}}}{\mathbf{G^{MWF}}^\dagger \boldsymbol{R}_N \mathbf{G^{MWF}}}. \tag{3.9}$$

By using equation (2.53) to exploit the decomposition of the MWF, we obtain

$$\begin{aligned}
\gamma^{out}(\nu) &= \frac{(G^{WF}\,\mathbf{G^{MVDR}}\,\tilde{H}_d^*)^\dagger \boldsymbol{R}_S (G^{WF}\,\mathbf{G^{MVDR}}\,\tilde{H}_d^*)}{(G^{WF}\,\mathbf{G^{MVDR}}\,\tilde{H}_d^*)^\dagger \boldsymbol{R}_N (G^{WF}\,\mathbf{G^{MVDR}}\,\tilde{H}_d^*)} \\
&= \frac{|G^{WF}|^2 |\tilde{H}_d|^2 \mathbf{G^{MVDR}}^\dagger \boldsymbol{R}_S \mathbf{G^{MVDR}}}{|G^{WF}|^2 |\tilde{H}_d|^2 \mathbf{G^{MVDR}}^\dagger \boldsymbol{R}_N \mathbf{G^{MVDR}}} \\
&= \frac{\mathbf{G^{MVDR}}^\dagger \boldsymbol{R}_S \mathbf{G^{MVDR}}}{\mathbf{G^{MVDR}}^\dagger \boldsymbol{R}_N \mathbf{G^{MVDR}}}.
\end{aligned} \tag{3.10}$$

Since $\tilde{H}_d$ is canceled out in the nominator and denominator term, the narrow-band output SNR of the MWF is independent of the particular choice of $\tilde{H}_d$.

3.2.2 Broad-band SNR Considerations

Next, the broad-band output SNR χ^{out} of the MWF is considered, which is defined as

$$\begin{aligned}
\chi^{out} &= \frac{\sum_\nu \mathbb{E}\left\{|Z_S(\nu)|^2\right\}}{\sum_\nu \mathbb{E}\left\{|Z_N(\nu)|^2\right\}} \\
&= \frac{\sum_\nu \mathbf{G^{MWF}}(\nu)^\dagger \boldsymbol{R}_S(\nu) \mathbf{G^{MWF}}(\nu)}{\sum_\nu \mathbf{G^{MWF}}(\nu)^\dagger \boldsymbol{R}_N(\nu) \mathbf{G^{MWF}}(\nu)}.
\end{aligned} \tag{3.11}$$

Since the MVDR beamformer achieves a distortionless response, the PSD of the speech component at its output is Φ_X^2. Hence, the PSD of the speech component Z_S at the output of the MWF can be written as

$$\mathbb{E}\left\{|Z_S(\nu)|^2\right\} = |G^{WF}|^2 |\tilde{H}_d|^2 \Phi_X^2 \,. \tag{3.12}$$

By using the noise PSD at the output of the MVDR beamformer from equation (2.56), the PSD of the noise component at the output of the MWF equals

$$\mathbb{E}\left\{|Z_N(\nu)|^2\right\} = |G^{WF}|^2 |\tilde{H}_d|^2 \Phi_{N,MVDR}^2 \tag{3.13}$$

and therefore the broad-band output SNR

$$\chi^{out} = \frac{\sum_\nu \Phi_X^2(\nu) |G^{WF}(\nu)|^2 |\tilde{H}_d(\nu)|^2}{\sum_\nu \Phi_{N,MVDR}^2(\nu) |G^{WF}(\nu)|^2 |\tilde{H}_d(\nu)|^2} \,. \tag{3.14}$$

From this equation, it can be seen that the overall transfer function as well as the single-channel Wiener post filter impact the broadband output SNR.

3.2.3 Broad-band SNR Dependence on the Overall Transfer Function

Next, the response $\tilde{H}_d$ that maximizes the broadband output SNR is considered. Equation (3.14) can be rewritten as

$$\chi^{out} = \frac{\sum_\nu \alpha_\nu |\tilde{H}_d(\nu)|^2}{\sum_\nu \beta_\nu |\tilde{H}_d(\nu)|^2} = \frac{\tilde{\mathbf{H}}_\mathbf{d}^\dagger \mathbf{A} \tilde{\mathbf{H}}_\mathbf{d}}{\tilde{\mathbf{H}}_\mathbf{d}^\dagger \mathbf{B} \tilde{\mathbf{H}}_\mathbf{d}} \tag{3.15}$$

with

$$\begin{aligned}
\alpha_\nu &= \Phi_X^2(\nu)|G^{WF}(\nu)|^2 \\
\beta_\nu &= \Phi_{N,MVDR}^2(\nu)|G^{WF}(\nu)|^2 \\
\tilde{\mathbf{H}}_\mathbf{d} &= [\tilde{H}_d(0), \dots, \tilde{H}_d(L-1)]^T \\
\mathbf{A} &= \begin{pmatrix} \alpha_0 & 0 & \dots & 0 \\ 0 & \alpha_1 & \dots & 0 \\ 0 & \dots & \ddots & 0 \\ 0 & \dots & 0 & \alpha_{L-1} \end{pmatrix} \\
\mathbf{B} &= \begin{pmatrix} \beta_0 & 0 & \dots & 0 \\ 0 & \beta_1 & \dots & 0 \\ 0 & \dots & \ddots & 0 \\ 0 & \dots & 0 & \beta_{L-1} \end{pmatrix},
\end{aligned}$$

where L denotes the total number of frequency bins. Maximizing χ^{out} is equivalent to solving the generalized eigenvalue problem $\mathbf{A}\tilde{\mathbf{H}}_\mathbf{d} = \lambda\mathbf{B}\tilde{\mathbf{H}}_\mathbf{d}$ or $\mathbf{B}^{-1}\mathbf{A}\tilde{\mathbf{H}}_\mathbf{d} = \lambda\tilde{\mathbf{H}}_\mathbf{d}$. The solution to the eigenvalue problem is the eigenvector corresponding to the largest eigenvalue λ_{max}. Since $\mathbf{B}^{-1}\mathbf{A}$ is a diagonal matrix, the largest eigenvalue is

$$\lambda_{max} = \max_\nu \frac{\alpha_\nu}{\beta_\nu} = \max_\nu \frac{\Phi_X^2(\nu)}{\Phi_{N,MVDR}^2(\nu)} . \tag{3.16}$$

Comparing equation (3.16) with (3.14), we obtain the corresponding eigenvector $\tilde{\mathbf{H}}_\mathbf{d} = [0, \dots, 1, \dots, 0]^T$ with a one in the frequency bin corresponding to the largest eigenvalue and zero elsewhere. Although this overall transfer function maximizes the broadband output SNR, the corresponding speech distortion will not be acceptable because only one frequency bin will pass the beamformer.

As an important conclusion, no optimal $\tilde{H}_d$ exists to fulfill both requirements due to this contradiction. The particular choice of $\tilde{H}_d$ has no effect on the narrow-band SNR and therefore still achieves optimal narrow-band noise reduction. However, the proper selection of $\tilde{H}_d$ allows a broad-band SNR improvement without increasing speech distortion compared with a single microphone signal. In the following chapter of this thesis, directivity based reference choices for $\tilde{H}_d$ are investigated by additionally taking the phase between the microphone signals into account. This allows to exploit the properties of the spatial noise field in the design process of a suitable $\tilde{H}_d$.

3.3 Summary

In this chapter, the generalized multichannel Wiener filter was derived. It was shown that the standard multichannel Wiener filter is a special case thereof, where the overall transfer function equals a dedicated microphone reference channel. However, the generalized multichannel Wiener filter expands this approach by applying an appropriate weighting of the individual channels in terms of a suitable reference vector, which allows to design the resulting overall transfer function. By the distinction between the narrow-band and the broad-band SNR, it could be shown that the narrow-band SNR is independent of the choice of the overall transfer function. However, the choice affects the broad-band SNR. The theoretically optimal solution for the broad-band SNR is not acceptable regarding the speech distortion. This raises the need for a suitable design of the overall transfer function, which will be shown in the following chapter.

4 Directivity Based Reference for the Generalized Multichannel Wiener Filter

In this chapter, different references for the G-MWF are presented. In [6], the magnitude of the response $\tilde{H}_d$ was designed to improve the broadband output SNR, whereas the phase term of $\tilde{H}_d$ was set equal to the phase of the ATF in the reference microphone. In [7], an MWF formulation with partial equalization (P-MWF) was introduced, where the overall transfer function was chosen as the envelope of the individual ATFs with the phase of an arbitrary reference microphone. This results in a partial equalization of the acoustic system and an improved broadband output SNR. While this approach has advantages with respect to background noise reduction, the reverberation caused by the acoustic environment is not reduced. As shown in [40], the all-pass component of a room impulse response contains strong reflexions and reverberation. This suggests that the phase reference of the G-MWF should be properly designed in order to improve the output signal-to-reverberation ratio (SRR). Concepts that aim to reduce reverberation by a suitable combining of the microphone signals were proposed in [41, 42, 43] and many others.

In contrast to the approaches in [6, 7], in this work a complex-valued selection vector $\mathbf{u} \in \mathbb{C}$ is considered, which takes the phase into account for the design of a directivity based overall transfer function $\tilde{H}_d$. Firstly, a spatially distributed microphone arrangement is considered. Two references are presented that can improve the signal-to-reverberation ratio (SRR) and the broadband output SNR compared with the S-MWF and the P-MWF. Both proposed references are based on a delay-and-sum beamformer (DSB), which aims to augment the direct path of arrival of the speech source. One approach uses the delay-and-sum beamformer directly as a reference, whereas the other approach combines the phase of the DSB with the reference of the P-MWF.

Further directivity based references for the G-MWF are introduced in this chapter which are based on differential microphone array beamforming [9]. There-

S. Grimm, *Directivity Based Multichannel Audio Signal Processing For Microphones in Noisy Acoustic Environments*, Schriftenreihe der Institute für Systemdynamik (IDS) und optische Systeme (ISO), https://doi.org/10.1007/978-3-658-25152-9_4

fore a closely spaced microphone arrangement in endfire configuration is considered to form a bipolar, a cardioid and a hypercardioid beam pattern. This allows to create a directional response for the desired overall transfer function $\tilde{H}_d$. As a result, the broadband output SNR as well as the log spectral distance (LSD) can be potentially improved, due to the suppression of noise and reflections coming from other incident angles than the direction of the speech source.

For these directivity based references, knowledge of the second order statistics of the speech and noise signals as well as the time-difference-of-arrival (TDOA) of the speech source between the microphones is required. In the literature several methods for TDOA-estimation were proposed [44, 45, 46, 47, 48, 49, 50]. Many of these techniques are summarized in [51]. The TDOA estimation is closely related to the estimation of the linear phase term of the relative transfer function (RTF) between the speech signals. Using the approach proposed in [52], which allows to obtain an unbiased RTF estimate in noisy conditions, an unbiased TDOA-estimate can be obtained from the non-diagonal elements of the speech correlation matrix $\boldsymbol{R}_S$.

This chapter is outlined as follows. The design of the overall transfer function for some special cases of the G-MWF are presented in section 4.1. In section 4.2, the additional knowledge of the TDOA between the signals is used to create the directivity based reference vectors. By applying the delay-and-sum approach to form a reference, reverberation is reduced due to coherent combining of the direct signal path. Differential beamforming is used to form a reference, which creates a spatial beam pattern for the overall transfer function. The block diagram of the system is presented in section 4.3. In section 4.4, the estimation of the noise and speech correlation matrices in stationary noise conditions is presented, while the related TDOA estimation and the challenge of acquiring these estimates in noisy and reverberated environments due to biased estimates is presented in section 4.5. In section 4.6, simulation results are presented. The delay-and-sum references are applied for two different simulation scenarios with distributed microphones and the SNR and SRR improvements are examined. The differential beamforming references are applied in the context of a monaural hearing aid and closely spaced microphones in two different environments and the results regarding the SNR improvement as well as the LSD are presented. Finally a summary is given in section 4.7.

The content of this chapter has been partly published in [11, 10] and [12].

4.1 G-MWF Reference Selection

In the following, special cases for the design of the G-MWF reference vector are presented. Note that the different formulations of the G-MWF differ only with respect to the vector $\mathbf{u}$ and the corresponding transfer function $\tilde{H}_d$. The resulting overall transfer functions only equal $\tilde{H}_d$ if no additional post filtering is applied ($\mu = 0$).

4.1.1 Distortionless Response Beamformer

The distortionless response beamformer obtains perfect equalization of the acoustic system, where the overall transfer function is chosen to be $\tilde{H}_d = 1$. Hence, the elements of the reference vector $\mathbf{u}$ are

$$u_i = \frac{H_i}{\mathbf{H}^\dagger \mathbf{H}} . \tag{4.1}$$

However, the resulting G-MWF requires perfect knowledge of the ATFs from the speaker to the microphones, which is often not available in practice.

4.1.2 Partial Equalization Approach

In [7] the P-MWF has been presented, where the amplitude of the overall transfer function is defined as the envelope of the individual ATFs, and the phase is chosen as the phase ϕ_{ref} of an arbitrary (reference) ATF, i.e.,

$$\tilde{H}_d = \sqrt{\mathbf{H}^\dagger \mathbf{H}}\, e^{j\phi_{ref}} . \tag{4.2}$$

This formulation results in a partial equalization of the acoustic system, since the dips in the magnitude response of the individual ATFs can be avoided. The elements of the vector $\mathbf{u}$ can be computed as

$$u_i = \sqrt{\frac{r_{S_{i,i}}}{\mathrm{tr}(\boldsymbol{R}_S)}} \frac{r_{S_{i,ref}}}{|r_{S_{i,ref}}|} = \frac{H_i}{\sqrt{\mathbf{H}^\dagger \mathbf{H}}} e^{-j\phi_{ref}} , \tag{4.3}$$

where again $\mathrm{tr}(\cdot)$ denotes the trace of the matrix and $r_{S_{i,j}}$ denotes the element of $\boldsymbol{R}_S$ in the ith row and j^{th} column. Hence, for the P-MWF we have

$$\boldsymbol{R}_S \mathbf{u} = \boldsymbol{R}_S \frac{\mathbf{H}}{\sqrt{\mathbf{H}^\dagger \mathbf{H}}} e^{-j\phi_{ref}} = \Phi_X^2 \mathbf{H} \sqrt{\mathbf{H}^\dagger \mathbf{H}}\, e^{-j\phi_{ref}} . \tag{4.4}$$

Compared to the MVDR beamformer reference in section 4.1.1, the advantage of the P-MWF is that it only depends on estimates of the second order statistics, i.e. $\boldsymbol{R}_S$ and $\boldsymbol{R}_N$. No explicit knowledge of the ATFs is required. Compared to the S-MWF, which also only depends on the signal statistics, the P-MWF approach is able to partially equalize the magnitude response of the ATFs. However, it should be noted that the output signal is as reverberant as the selected reference ATF due to the reference channel selection of the phase component.

4.2 Directivity Based G-MWF Reference Selection

In the following, the new reference selection approaches for the G-MWF are introduced, which take the knowledge of the TDOA into account to form a directivity based overall transfer function. Two references for spatially distributed microphones in a reverberant environment are introduced, which use a phase reference from a DSB to augment the direct path of the desired signal. Furthermore, for the application of closely spaced microphone arrays, differential beamforming references are presented, which allow to form a directional response for the overall transfer function.

The choice of the reference selection is dependent on the microphone arrangement, the location of the desired signal source and the acoustic environment. The DSB references are suitable for microphone arrangements with larger distances between the acoustic sensors to benefit from the spatial diversity of the sound field. For closely spaced sensors, the acoustic transfer functions are nearly identical, so the spatial sampling of the sound field is very limited and no significant improvement regarding SNR and SRR can be made by this approach compared with a single microphone. However, for the differential reference choices, which rely on sound pressure differentials between the microphones, the sensor spacing has an impact on the performance. By placing the sensors in greater distance, acoustic influences and decorrelation of the signals (especially for the noise terms) violate the assumptions necessary for the differential beamforming and degrade the noise reduction capability. As a result, the microphones need to be spaced closely in contrast to the DSB reference choices.

4.2.1 Delay-and-Sum Beamformer

In the first approach it is proposed to simply use the output of a delay-and-sum beamformer as the speech reference. The corresponding elements of the vector $\mathbf{u}$ can be described as

$$u_i = \frac{1}{M} \cdot e^{j2\pi\nu\tau_i}, \tag{4.5}$$

where τ_i is a delay, which compensates the TDOA of the direct path speech components at the microphones. The speech components are typically aligned in relation to the microphone with the latest arrival time to obtain a causal DSB. Using (3.3), the overall transfer function is obtained

$$\tilde{H}_d = \frac{1}{M} \sum_i H_i e^{-j2\pi\nu\tau_i}. \tag{4.6}$$

4.2.2 Partial Equalization with DSB Phase Reference

The second approach is a combination of the P-MWF of [7] with the DSB as the phase reference only. As already described in section 4.1.2, the phase reference of the P-MWF is the phase of an arbitrary ATF. In order to improve the SRR, the DSB can be used as the phase reference. The resulting vector $\mathbf{u}$ can be described as

$$u_i = \sqrt{\frac{r_{S_{i,i}}}{\mathrm{tr}(\boldsymbol{R}_S)}} \cdot e^{j2\pi\nu\tau_i}\,, \tag{4.7}$$

where again τ_i is a delay to compensate the TDOA of the speech signal direct path. Note that the phase term impacts the magnitude of the overall transfer function $\tilde{H}_d$, cf. (3.3). Comparing (4.3) and (4.7) results in

$$u_i = \frac{|H_i|}{\sqrt{\mathbf{H}^\dagger \mathbf{H}}} e^{j2\pi\nu\tau_i} \tag{4.8}$$

and the overall transfer function equals

$$\tilde{H}_d = \frac{1}{\sqrt{\mathbf{H}^\dagger \mathbf{H}}} \sum_i |H_i| H_i e^{-j2\pi\nu\tau_i} \tag{4.9}$$

$$= \frac{1}{\sqrt{\mathbf{H}^\dagger \mathbf{H}}} \sum_i |H_i|^2 e^{j(\phi_i - 2\pi\nu\tau_i)}\,. \tag{4.10}$$

Hence, the direct path speech components in the microphones are aligned but additionally the microphone signals are weighted with the magnitude of the ATFs similar to the P-MWF approach. This achieves a partially equalized magnitude response for the overall transfer function due to the magnitude weighting, while simultaneously the DSB phase reference improves the dereverberation capabilities of the chosen reference vector design.

4.2.3 Differential Beamforming Reference

The differential beamforming references allow to form a spatial beam pattern as an overall transfer function $\tilde{H}_d$ for the G-MWF. Similar to the delay-and-sum beamformer, the TDOA of the speech signal at the microphones is required to form the directional response. In contrast to the DSB, the microphones need to be positioned in an endfire array configuration pointing towards the desired speech source to achieve their best performance. The described references for the G-MWF form first order differential array beam patterns, which means a minimum of two microphones is required to realize the desired directional response [9]. The described references only differ in the delay τ to form the required beam pattern. In this work, an endfire array consisting of two microphones is considered for the differential beamforming references.

Dipole Pattern

The dipole pattern reference creates a null for signals arriving from an incident angle of $90°(270°)$ regarding the position of the endfire array. No delay elements are required to realize this reference, which creates an $\tilde{H}_d$ with the desired directional behavior. For an array consisting of two microphones, the corresponding reference vector $\mathbf{u}$ can be written as

$$\mathbf{u} = G^{EQ^*}[1, -1]^T , \tag{4.11}$$

where G^{EQ} denotes a compensation filter for the first order high pass behavior created by the differential output.

Cardioid Pattern

The cardioid pattern reference is able to create a single null at an incident angle of $180°$ for the overall transfer function $\tilde{H}_d$, which means signals coming directly from the back of the endfire array are totally suppressed. τ_0 is the propagation time of planar sound waves traveling between the two microphones as derived in (2.22). For $M = 2$ microphones, $\mathbf{u}$ equals to

$$\mathbf{u} = G^{EQ^*}[1, -e^{j2\pi\nu\tau_0}]^T . \tag{4.12}$$

Hypercardioid Pattern

The hypercardioid beam pattern reference creates a null at an incident angle of $110°(250°)$. In sound fields consisting of diffuse noise, the hypercardioid beam pattern is able to achieve the biggest SNR improvement due to its directivity index. The corresponding reference vector $\mathbf{u}$ for the G-MWF is realized as

$$\mathbf{u} = G^{EQ^*}[1, -e^{j2\pi\nu\frac{\tau_0}{3}}]^T . \tag{4.13}$$

Equalization Filter G^{EQ}

G^{EQ} is the required equalization filter to compensate the high pass behavior of the differential microphone array output. Additionally, the gain loss needs to be compensated with a constant amplification factor. By comparing the differential array references for the G-MWF with (2.21), G^{EQ} can be written as

$$G^{EQ} = \frac{C}{\nu}, \tag{4.14}$$

where the constant C compensates for the gain loss and $\frac{1}{\nu}$ equalizes the high pass behavior due to the differential output characteristic.

As an alternative approach, the filter G^{EQ} can be derived in anechoic and noise free conditions in the MMSE sense by minimizing the error between the uncompensated output of the differential array and a selected reference speech signal

$$G^{EQ} = \underset{G^{EQ}}{\operatorname{argmin}}\, \mathbb{E}\left\{|([1, -e^{-j2\pi\nu\tau}]\mathbf{S})G^{EQ} - S_{ref}|^2\right\}. \tag{4.15}$$

4.3 System Structure of the G-MWF

Figure 4.1 depicts the block diagram of the G-MWF for an array of two microphones. Since the filtering is performed in the frequency domain, the microphone signals are first windowed and then transformed using the fast Fourier transform (FFT).

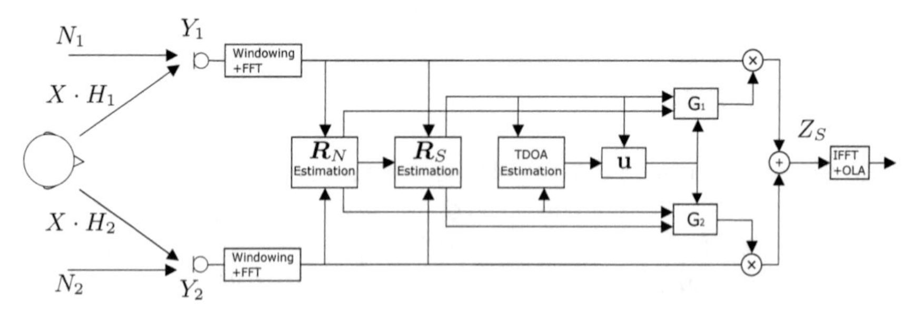

Figure 4.1: System structure for multichannel Wiener filtering with vector $\mathbf{u}$ [10]

4.4 Estimation of the Second Order Statistics for Speech and Noise

One important aspect of noise reduction filtering is the estimation of the speech and noise correlation matrices $\boldsymbol{R}_S$ and $\boldsymbol{R}_N$, since the actual PSDs and cross power spectral densities (CPSDs) are not available in practice. Often a voice activity detector (VAD) [53, 54] is used to determine whether speech is present or absent to update the noise or speech PSD estimates. Another approach to obtain these estimates is based on minimum statistics [55, 56] with the advantage that no VAD is required. A frequency-dependent VAD, as proposed in [57], allows to update the noise estimates even during speech presence, since speech is a signal that is sparse in time as well as in the frequency domain. The speech activity detection in [57] is based on a simple threshold test, where the current input power is compared to the estimated noise PSD which is multiplied by a threshold value

$$VAD(\eta,\nu) = \begin{cases} 1, & \text{for } |Y(\eta,\nu)|^2 \geq \Phi_N^2(\eta,\nu)\Theta(\nu) \\ 0, & \text{else}. \end{cases} \tag{4.16}$$

$\Theta(\nu)$ denotes the threshold value which is allowed to be set frequency dependent (for more details, see [57]). In case speech is absent, i. e. $VAD(\eta,\nu) = 0$, $\boldsymbol{R}_N$ is updated by the noise PSD and CPSD estimates. These are obtained by recursive smoothing in frequency bins where speech is absent

$$r_{N_{i,j}}(\eta,\nu) = \begin{cases} \zeta r_{N_{i,j}}(\eta-1,\nu) + (1-\zeta)Y_i(\eta,\nu)Y_j(\eta,\nu)^*, & \text{for } VAD(\eta,\nu) < 1 \\ r_{N_{i,j}}(\eta-1,\nu), & \text{else}. \end{cases} \tag{4.17}$$

ζ denotes the recursive smoothing parameter. Since speech and noise are assumed to be uncorrelated, the estimate of the speech correlation matrix can be obtained by

$$\boldsymbol{R}_S = \boldsymbol{R}_Y - \boldsymbol{R}_N\,. \tag{4.18}$$

4.5 TDOA Estimation

For the directivity based reference vectors as proposed in section 4.1, the TDOA from the speaker to the microphones is required to achieve a coherent summation of the direct path for the delay and sum beamformer, as well as to steer the spatial nulls of the differential beamformer reference. A very popular TDOA estimation approach is the generalized cross correlation (GCC) method [44, 49, 51], where the cross-correlation between the microphone signals is calculated in the frequency domain as the cross power spectral density. Depending on the application and the environmental conditions, the CPSD is typically filtered by a weighting function to achieve a better TDOA estimation in terms of a sharper peak of the cross-correlation in the time domain. For example, the GGC-PHAT method performs a spectral whitening of the CPSD magnitude spectrum to obtain a CPSD which only depends on the phase relation between the signals. This approach achieves good estimates in reverberant environments [47], whereas for noisy conditions a coherence-based or SNR-based weighting function is preferred [58].

To obtain the TDOA, the weighted CPSD is transformed to the time domain using the inverse Fourier transform resulting in the cross correlation vector. The main peak in the cross correlation vector indicates the time delay in samples. To achieve a sub-sample TDOA estimate in terms of fractional delays, methods as the parabolic fitting can be used to acquire a more accurate estimate [59, 60]. The TDOA estimate is only valid in signal blocks where the speaker is active, which can be determined by using voice activity detection.

It should be noted that the phase of the CPSD is equal to the phase of the relative transfer function (RTF) between the microphones since both only differ from a different magnitude response. Since in general the microphone signals contain correlated noise components, estimating the RTFs directly from the noisy microphone signals leads to biased RTF estimates. Several methods for unbiased RTF estimation have been proposed, e.g., by exploiting the non-stationarity of speech signals [38, 61] or by using the generalized eigenvalue decomposition of $\boldsymbol{R}_Y$ and $\boldsymbol{R}_N$ [62, 63]. In [52], an approach for unbiased RTF estimation was proposed, requiring estimates of the PSDs and CPSDs of the speech and noise components, which can be obtained from the estimated speech and noise correlation matrices $\boldsymbol{R}_S$ and $\boldsymbol{R}_N$. The RTF estimate between microphones i and j is computed as a combination of two weighted coefficients

$$\hat{W}_{unbiased} = f_i \frac{r_{S_{i,j}}}{r_{S_{i,i}}} + f_j \frac{r_{S_{j,j}}}{r_{S_{j,i}}}, \tag{4.19}$$

where the terms f_i and f_j are SNR based weighting coefficients which are defined as

$$f_i = \frac{\frac{r_{S_{i,i}}}{r_{N_{i,i}}}}{\frac{r_{S_{i,i}}}{r_{N_{i,i}}} + \frac{r_{S_{j,j}}}{r_{N_{j,j}}}} \tag{4.20}$$

$$f_j = \frac{\frac{r_{S_{j,j}}}{r_{N_{j,j}}}}{\frac{r_{S_{i,i}}}{r_{N_{i,i}}} + \frac{r_{S_{j,j}}}{r_{N_{j,j}}}} . \tag{4.21}$$

The approach in [52] is slightly modified, based on a frequency dependent VAD [57], where the RTF estimate is updated only in frequency bins where speech activity is detected. Furthermore, a recursive smoothing parameter to average the RTF estimate is used, which is the rate of all frequency bins where speech activity is detected. By applying the inverse Fourier transform, $\hat{W}_{unbiased}$ can be transformed back into the time domain, which results in the vector $\hat{w}_{unbiased}$. The location of the peak value that indicates the delay to the microphone j can be calculated as

$$\tau_i = \max_{k=0,\dots,L-1} \hat{w}_{unbiased}(k) , \tag{4.22}$$

where $\hat{w}_{unbiased}(k)$ is the k-th element of the vector $\hat{w}_{unbiased}$.

4.6 Simulation Results

In the following, the simulation results for the proposed references of the G-MWF are presented. Since the performance of the different G-MWF approaches are dependent on the choice of the microphone arrangement, simulation environments with different arrangements are chosen, based on the reference selection. For the DSB references, simulation environments with distributed microphones are investigated, which offer the potential to acquire spatial diversity. One of them is a vehicle interior, since acoustically it offers a very short reverberation time but has a low SNR due to the driving noise. The other one is a classroom, which has a good SNR but is very reverberant.

For the differential beamforming approaches, a closely spaced microphone arrangement of a monaural hearing aid is examined since this offers the potential to form directional beam patterns. Simulations are carried out in a rather anechoic space as well as a more realistic scenario which allows to investigate the influence of reverberant acoustics on the performance.

4.6.1 Delay-and-Sum Reference

The Simulation Environment

To investigate the potential SRR and SNR improvements provided by the proposed DSB reference for the G-MWF, simulations were carried out in two different environments as depicted in Figure 4.2. One is a noisy car environment, which could be the use case of a typical hands-free communications situation (Figure 4.2a). The other scenario is a reverberant classroom, which could reflect a teleconferencing situation (Figure 4.2b). All simulations were performed with a sampling rate of 16 kHz and an FFT length $L = 512$ with an FFT shift of 128 samples. For the overlapp-add implementation a Hamming window was used. The signals for testing the algorithms are ITU speech signals convolved with measured impulse responses. For the car scenario, this was done with an artificial head and two cardioid microphones that were mounted close to the rear-view mirror. For the class-room scenario, the acoustic measurements in [64] were used. These include impulse responses which were recorded with a loudspeaker and omnidirectional microphones at two different spatial locations with a microphone distance of 0.5 m. The reverberation time RT_{60} of the class-room has a value between 1.5 and 1.8 seconds over all frequencies.

Energy Decay Curves

In the following, G-MWF-1 denotes the G-MWF that uses the DSB as the speech reference, i.e. (4.5), whereas G-MWF-2 denotes the partial equalization approach, using the DSB only as a phase reference, i.e. (4.7). For the S-MWF and the P-MWF, Y_1 was used as the reference. To evaluate the dereverberation capabilities of the algorithms, the energy decay curves (EDCs) [65] of the resulting overall transfer functions $\tilde{H}_d$ were calculated (for $\mu = 0$) using the measured impulse responses. The resulting EDCs are shown in Figure 4.3.

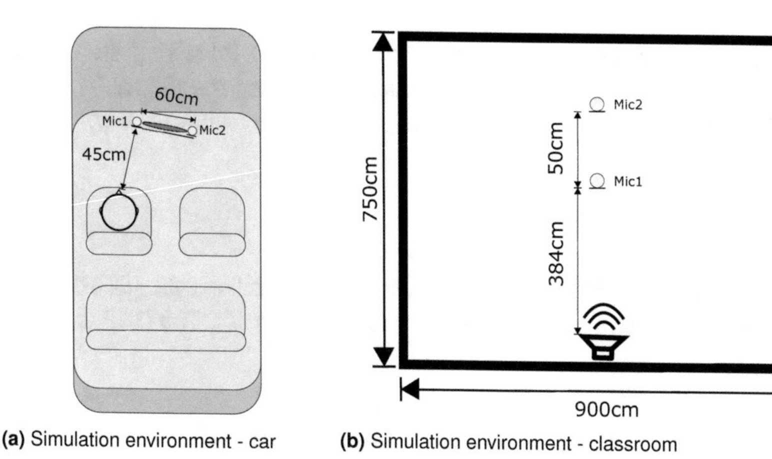

(a) Simulation environment - car

(b) Simulation environment - classroom

Figure 4.2: The two different simulation environments

For the car environment, the resulting EDCs are shown in Figure 4.3a. Curve (a) depicts the EDC of the overall transfer function for the S-MWF. Curve (b) depicts the resulting EDC of the overall transfer function of the P-MWF. Compared with (a), it can be observed that the decay time is increased but the energy of the first reflections is reduced due to the partial equalization as can be seen from the first 230 samples of the EDC. Curve (c) and (d) depict the EDC of the overall transfer function for the G-MWF-1 and G-MWF-2 respectively. Compared with (a) and (b), a reduced decay time is observed due to the coherent combining of the phase terms. As a result, the direct components of the acoustic transfer functions are enhanced, which leads to an improvement in speech quality of the overall system.

For the class-room scenario, the resulting EDCs are shown in Figure 4.3b. Due to the longer reverberation time, compared with the car environment, the resulting EDCs show a different behaviour. Curve (e) and (f) depict the EDCs of the resulting transfer function for the S-MWF and the P-MWF respectively. Curve (g) and (h) depict the EDCs of the overall transfer functions for the G-MWF-1 and the G-MWF-2. Compared to (e) it can be observed in (f) that the direct signal component for the first few samples is augmented due to the partial equalization, but that the decay time is increased. While (h) still shows a slightly better performance than (g) for the first 7000 samples, the decay time is increased by a small amount compared with (h) during the samples $7000 - 10000$. However, the reverberation energy for the G-MWF-1 and G-MWF-2 in (g) and (h) is noticeably reduced compared with (e) and (f).

Direct-to-Reverberation Ratio

As a measure of reverberation, the direct-to-reverberation ratio (DRR) can be calculated from the resulting overall transfer functions $\tilde{H}_d$. The DRR is defined as [66]

$$DRR = 10\log_{10}\left(\frac{\sum_{k=0}^{k_d} \tilde{h}_d^2(k)}{\sum_{k=k_d+1}^{\infty} \tilde{h}_d^2(k)}\right) dB\,, \tag{4.23}$$

where $\tilde{h}_d$ is the impulse response of the overall transfer function $\tilde{H}_d$ in the time domain and k_d are the samples of the direct path. For k_d, a time interval of 8 ms after the first arrival of the direct sound is considered. In Table 4.1 the DRR values for the different overall transfer functions $\tilde{H}_d$ are presented. From the Table it can be seen that the G-MWF approaches improve the DRR in both scenarios compared with the S-MWF and P-MWF.

Table 4.1: DRR of the overall transfer function for choosing a different phase and magnitude reference

	S-MWF	P-MWF	G-MWF1	G-MWF2
car scenario	12.6 dB	9.3 dB	14.7 dB	14.3 dB
class-room scenario	-3.8 dB	-3.8 dB	-1.4 dB	-1.6 dB

Resulting Overall Transfer Function

Both versions of the G-MWF result in similar overall transfer functions. This can be observed in Figure 4.4. Figure 4.4a presents the magnitude response of the acoustic transfer functions H_1 and H_2 of the car environment for both microphones as well as the overall transfer function of G-MWF-2 for frequencies

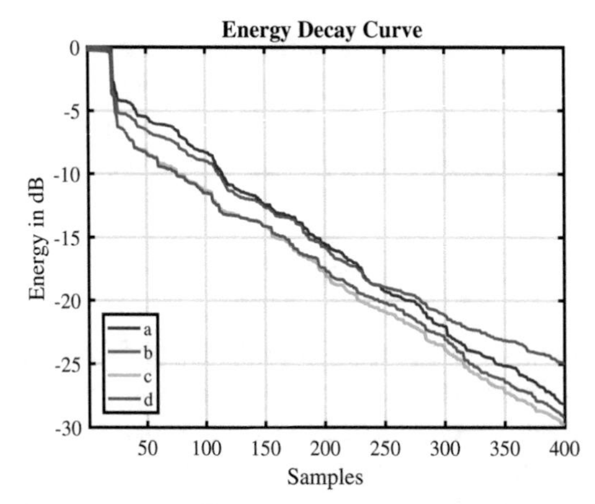

(a) EDC of the resulting acoustic transfer functions of the car environment: (a) ATF from the speech signal source to microphone 1 (S-MWF), (b) overall transfer function of P-MWF with phase reference of microphone 1, (c) overall transfer function of G-MWF-1 (d) overall transfer function of G-MWF-2 [10]

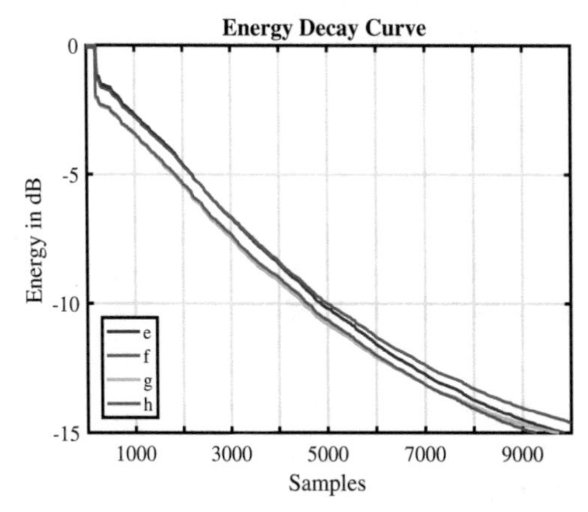

(b) EDC of the resulting acoustic transfer functions of the classroom environment: (e) ATF from the speech signal source to microphone 1 (S-MWF), (f) overall transfer function of P-MWF with phase reference of microphone 1, (g) overall transfer function of G-MWF-1 (h) overall transfer function of G-MWF-2 [10]

Figure 4.3: Energy decay curve for different $\tilde{H}_d$ formulations in the two simulation environments

between 2600 and 4000 Hz. The resulting partial equalization of the G-MWF-2 can be clearly seen. Figure 4.4b depicts the overall transfer function of both G-MWF versions for the same frequency section but with different scaling. It is shown that the magnitude responses of both approaches are quite similar.

Segmental Output SNR and SRR Improvement

Finally, we consider a noisy car scenario to evaluate the SNR and SRR improvement capabilities. The noise was recorded at a driving speed of 100 km/h with the same microphone set-up as specified in the simulation environment. For $\mu > 0$, the MWF performs an adaptive noise reduction with the estimated speech and noise correlation matrices as described in section 4.4 and therefore the resulting overall transfer function is time-varying. As a result, signal-based performance measures for the noise reduction and dereverberation performance need to be used. For the dereverberation performance, the signal-to-reverberation ratio (SRR) after [66] is used, i.e.,

$$SRR = 10\log_{10}\left(\frac{\mathbb{E}\left\{|s_d(k)|^2\right\}}{\mathbb{E}\left\{|z_s(k) - s_d(k)|^2\right\}}\right) dB\,, \tag{4.24}$$

where $s_d(k)$ is the direct path signal component of the reference microphone and $z_s(k)$ is the output signal of the algorithm in the time domain. It should be noted that this measure is only valid for signal segments where speech activity is detected.

Regarding the broadband output SNR improvement, the segmental output SNR (SSNR) is used for evaluation [67]. It is calculated as the average SNR value of speech active signal frames

$$SSNR = \frac{1}{K}\sum_{q=0}^{K-1}\left[10\log_{10}\left(\frac{\sum_{k=Pq}^{Pq+O-1}|\tilde{s}(k)|^2}{\sum_{k=Pq}^{Pq+O-1}|\tilde{n}(k)|^2}\right)\right]_{-10}^{35} \tag{4.25}$$

where K is the number of speech active frames and $\tilde{s}(k)$ and $\tilde{n}(k)$ are the speech and the noise components. O and P denote the frame length and the frame shift, which are chosen by 512 and 256 samples. An ideal voice activity detection was used to determine speech active frames and the maximum and

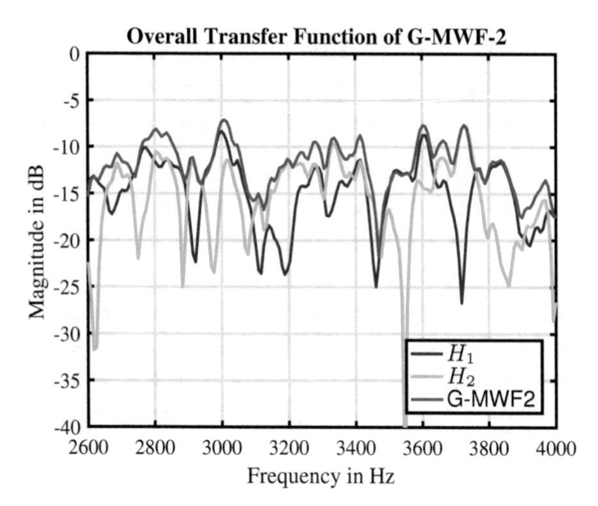

(a) Magnitude response of G-MWF2 [10]

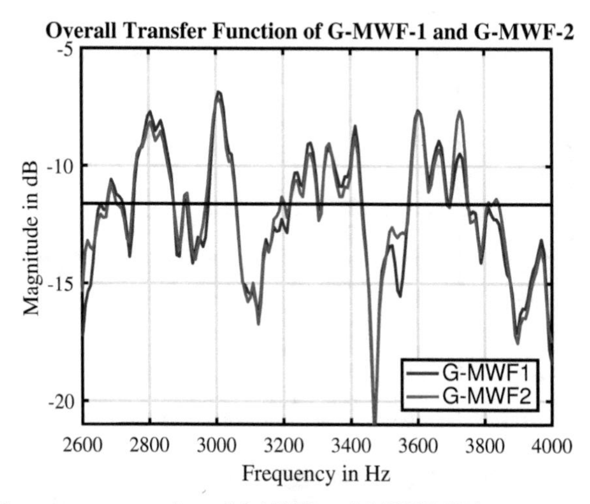

(b) Magnitude response comparison of G-MWF1 and G-MWF2 [10]

Figure 4.4: Magnitude response for different $\tilde{H}_d$ formulations

minimum SNR values for each frame are limited to -10 dB and 35 dB respectively. To evaluate the SSNR improvement, the speech and noise terms $z_s(k)$ and $z_n(k)$ at the output of the algorithm are used for the calculation.

Table 4.2 presents the results for the SRR and the segmental output SNR for two settings of the trade-off parameter μ, where a larger value of μ results in more noise reduction. The SRR was measured in time frames where speech was present. The performance of both G-MWF approaches are compared with the S-MWF and P-MWF. It can be observed that both G-MWF approaches outperform the S-MWF in terms of SRR and SSNR improvement. G-MWF-1 outperforms the P-MWF in terms of SRR and SSNR, whereas G-MWF-2 improves the SRR compared to G-MWF-1 at the expense of a small SSNR loss.

Table 4.2: SRR and SSNR comparison for different MWF formulations

$\mu = 0$	SSNR	SRR
S-MWF	-1.94 dB	2.87 dB
P-MWF	-0.86 dB	2.29 dB
G-MWF1	-0.72 dB	4.69 dB
G-MWF2	-1.33 dB	5.86 dB
$\mu = 30$	SSNR	SRR
S-MWF	2.82 dB	1.66 dB
P-MWF	4.35 dB	1.81 dB
G-MWF1	4.90 dB	3.49 dB
G-MWF2	4.25 dB	5.08 dB

As can be observed from the presented simulation results, the newly introduced reference choices are indeed capable to improve the SSNR while simultaneously reducing reverberation. This is achieved by taking the phase term into account for the design of the overall transfer function. Compared with the S-MWF and the P-MWF, the DSB references obtain an SSNR improvement as well as an augmentation of the direct signal for the car environment and the classroom scenario.

4.6.2 Differential Beamforming Reference

The Simulation Environment

For the simulations of the differential beamforming references, a simulation scenario with a monaural hearing aid consisting of two closely spaced microphones mounted on an artificial head is examined. Therefore an anechoic environment with an cylindrically isotropic noise field as well as a more realistic situation in a cafeteria is considered for the simulations. The datasets to create the simulations are obtained from the database of multichannel in-ear and behind-the-ear head-related and binaural room impulse responses in [68]. The sampling rate for all simulations is 48 kHz and a FFT length of $L = 512$ was used with a blockshift of 128 samples. For the overlapp-add implementation, a Hamming window was used. The spacing between the microphones of the hearing aid is 7.6 mm.

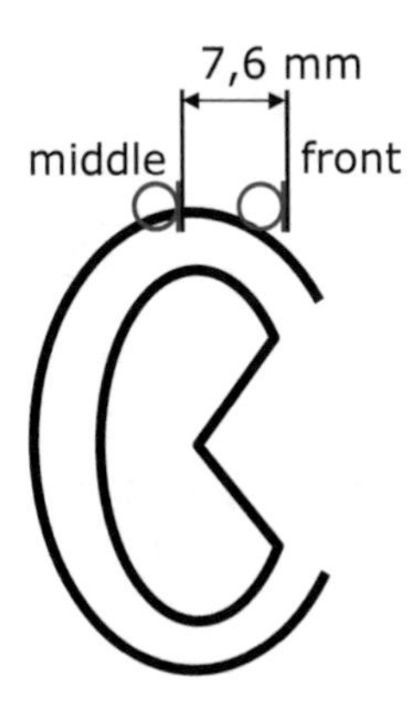

Figure 4.5: Microphone arrangement for the monaural hearing aid

For the anechoic environment as depicted in Figure 4.6a, the incident angle of the signal source can be varied in steps of $5°$. To create a cylindrically isotropic noise field, uncorrelated white Gaussian noise sources where convolved with the impulse responses from every possible incident angle of the signal source to the microphones of the hearing aid on the artificial head. For the cafeteria scenario as depicted in Figure 4.6b, a signal source with a distance of 102 cm and an incident angle of $0°$ was used for the simulations. The noise signals are background noise recorded with the microphones of the hearing aid. For both simulation scenarios, female and male ITU speech samples were used that were convolved with the impulse responses from the signal source to the microphones.

For both simulation scenarios, the PSDs, the CPSDs as well as the TDOA for the microphones are assumed to be known. The values of $\boldsymbol{R}_S$ and $\boldsymbol{R}_N$ are calculated for the whole dataset and the G-MWF is applied as a batch job.

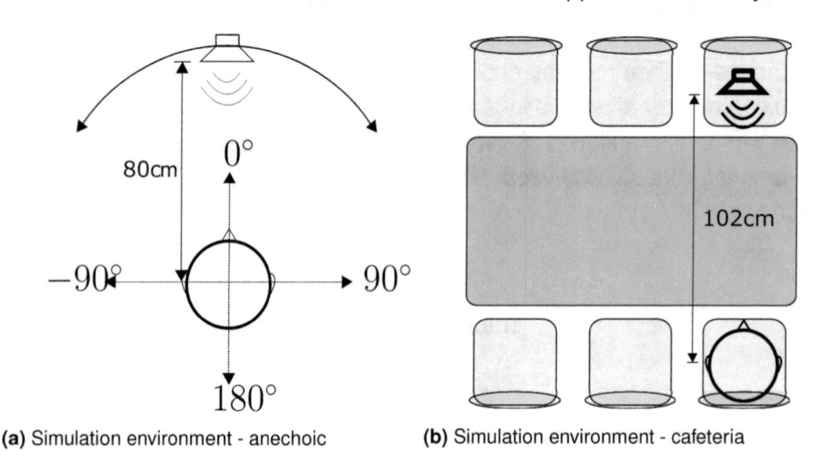

(a) Simulation environment - anechoic

(b) Simulation environment - cafeteria

Figure 4.6: The different simulation environments for the monaural hearing aid

Equalization Filter G^{EQ}

The equalization filter G^{EQ} for the differential beamforming references is calculated as described in (4.15) for the anechoic simulation environment and a signal source coming from an incident angle of $0°$, where the signal at the front microphone Y_1 is chosen as the reference. The resulting magnitude response of the equalization filter for the dipole, the cardioid as well as the hypercardioid is depicted in Figure 4.7.

As can be observed, the magnitude response of all filters look quite similar with a variation in the gain factor. The slope of the curves show a first order

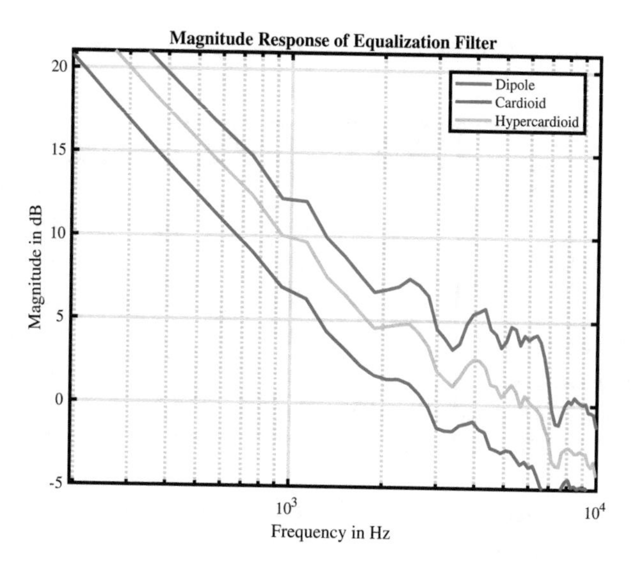

Figure 4.7: Magnitude response of equalization filter G^{EQ} for different beam patterns

low pass behavior to compensate for the first order high pass behavior of the differential output signal in the low frequencies, with only small derivations in the higher frequencies. The obtained filters are used in the following simulations for frequency response compensation on the anechoic as well as the cafeteria environment. Note that a compensation for very low frequencies does not make any sense due to the lack of speech signal energy.

Segmental SNR and LSD

In the following, the SSNR improvement capabilities of the proposed differential beamforming references compared with the S-MWF are examined. The SSNR is calculated for frequencies between 400 Hz and 4 kHz by band-limiting the signals. For further comparison, the P-MWF as well as the G-MWF1 (G-MWF with delay and sum reference) are also applied. The obtained results for the cylindrically isotropic noise field in the anechoic environment as well as the cafeteria environment with real background noise are presented in Table 4.3.

Table 4.3: Segmental SNR comparison for different G-MWF references (angle: $\theta = 0°$)

$\mu = 1$	Anechoic environment	Cafeteria environment
Microphone 1 (front)	5.6 dB	4.1 dB
S-MWF	7.7 dB	5.3 dB
G-MWF1	7.9 dB	5.2 dB
P-MWF	7.9 dB	5.2 dB
Dipole reference	8.5 dB	6.5 dB
Cardioid reference	8.6 dB	6.6 dB
Hypercardioid reference	8.8 dB	7.0 dB

As can be observed, the S-MWF is capable to improve the SSNR by over 2 dB for the anechoic environment. It should be noted that the SSNR gain is quite small in general due to the batch job processing, i.e., the filters are stationary. The P-MWF as well as the G-MWF1 perform quite similar to the S-MWF in both scenarios since the microphones are closely spaced and the channel diversity is quite small. However, the differential performing references show a significant SNR improvement for both simulation scenarios, where the hypercardiod reference selection shows the best performance regarding the SSNR.

For the cafeteria environment, a spectogram is shown in Figure 4.8. It depicts the output signal for the reference microphone, the S-MWF as well as the G-MWF with the hypercardioid reference. As can be seen, the S-MWF reduces the background signal energy, however, the G-MWF with the hypercardioid reference is able to reduce the background noise energy even more.

In Table 4.4 the results regarding the log spectral distance are presented. The log spectral distance measures the linear distortion in comparison to a reference spectrum. It is calculated as

$$LSD = \sqrt{\frac{1}{L}\sum_{\nu}\left[10\log_{10}\left\{\frac{\mathbb{E}\left\{|\tilde{S}(\nu)|^2\right\}}{\mathbb{E}\left\{|X(\nu)|^2\right\}}\right\}\right]^2} \tag{4.26}$$

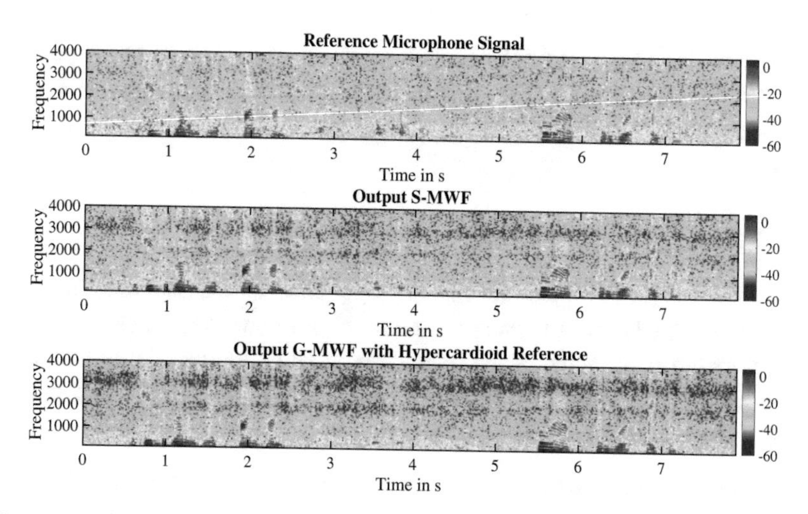

Figure 4.8: Spectogramm of the output signals for the cafeteria environment ($\mu = 3$)

Table 4.4: Log spectral distance comparison for different G-MWF references (angle: $\theta = 0°$)

$\mu = 1$	Anechoic environment	Cafeteria environment
Microphone 1 (front)	3.6 dB	3.9 dB
S-MWF	2.6 dB	4.5 dB
G-MWF1	2.3 dB	4.4 dB
P-MWF	2.4 dB	4.4 dB
Dipole reference	2.0 dB	4.5 dB
Cardioid reference	2.2 dB	4.4 dB
Hypercardioid reference	2.1 dB	4.4 dB

where $\tilde{S}(\nu)$ is the signal which is compared to the signal $X(\nu)$ at the mouth-reference point. For the LSD evaluation of the output signal of the algorithm $Z_S(\nu)$ is used for $\tilde{S}(\nu)$.

As can be observed, the values are slightly improved for the differential beamforming references in the anechoic scenario, while the P-MWF and the G-MWF1 show similar values compared with the S-MWF. In the cafeteria environment, the LSD for all references is similar to that of the S-MWF, however, the SNR improvement for the differential beamforming references is superior to the S-MWF.

Angle Dependent Segmental SNR

In the following, the segmental SNR for speech and noise sources of varying incident angles in the anechoic environment are examined for the differential beamforming references. Since only a monaural hearing aid for the left ear is used for the simulations, the angle is varied within $0°$ to $-180°$. The simulation results are depicted in Figure 4.9.

In Figure 4.9a, a single noise source of white Gaussian noise is varied for incident angles θ from $0°$ to $-180°$. The speech source stays at a fixed position of $\theta = 0°$. Compared with the S-MWF which uses the reference microphone, the dipole, the cardioid as well as the hypercardioid show an improved segmental SNR for nearly all incident angles except for incident angles between $-150°$ and $-180°$, where the dipole reference shows a slightly inferior performance. The SSNR has its maximum value mainly at the incident angles where the differential beamformer is supposed to suppress the noise source best. This is $-90°$ for the dipole, $-180°$ for the cardioid and $-110°$ for the hypercardioid beam pattern. As can be observed, these values are mainly identical to the simulation results.

In Figure 4.9b, a cylindrically isotropic noise field in the anechoic environment is considered. The speech source is varied within incident angles of $0°$ to $-180°$ and the G-MWF for the differential beamforming references is calculated. Compared with the S-MWF, which uses a reference microphone for the overall transfer function, the differential beamforming references show a superior SNR performance for incident angles of the speech source within $0°$ to $-40°$. If the speech source is coming from incident angles within $-45°$ to $-180°$, the signal is clearly suppressed compared with the S-MWF using a reference microphone. Again, the maximum suppression angles of the differential beamforming references can be detected by the SNR minimum values in the plot.

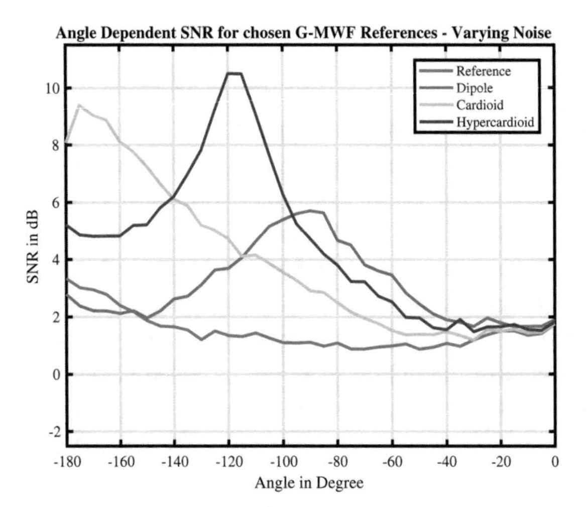

(a) Segmental SNR for an angle varying noise source and a fixed speech source (speech source angle: $\theta = 0°$) [12]

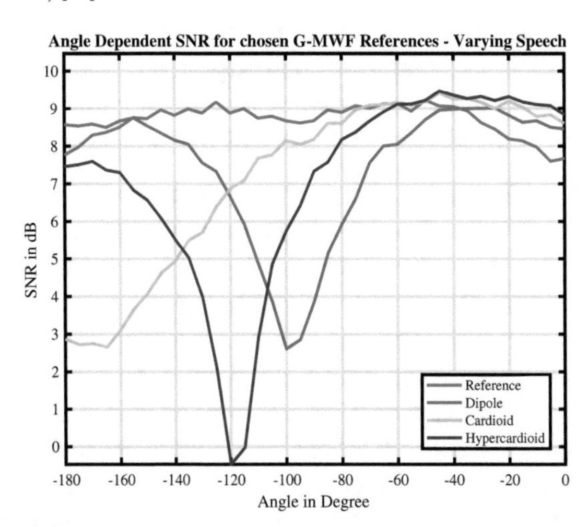

(b) Segmental SNR for an angle varying speech source in a cylindrical isotropic noise field (speech source angle: $\theta \in (0°, -180°)$

Figure 4.9: Angle dependent segmental SNR plots for directional $\tilde{H}_d$ references in anechoic environment

4.7 Summary

In this chapter, new references for the generalized multichannel Wiener filter were introduced. Instead of applying a weighted sum of the individual channels, additional information about the TDOA was used to design a class of directivity based references for the G-MWF. By applying the concept of delay-and-sum beamforming to augment the direct signal path of the desired speech source, two new references were introduced that aim to improve the broadband output SNR while simultaneously reducing reverberation. Therefore, the delay-and-sum beamfomer is used directly in one reference design. In another approach the delay-and-sum beamformer is combined with the P-MWF reference to obtain a partial equalization of the overall transfer function and reduced reverberation due to the DSB phase reference. By introducing a class of differential beamforming references for the G-MWF, the directional response is allowed to be implicitly designed. Also for this class of references the TDOA regarding the desired speech source is required to form a dipole, a cardioid and a hypercardioid G-MWF reference. Since the knowledge of the TDOA is essential for these directivity based references, a suitable estimate for noisy conditions is introduced based on an unbiased RTF estimation approach in [52].

To verify the reference choices, simulations were performed for two different microphone arrangements. For the delay-and-sum references, spatially distributed microphones were chosen for a noisy car as well as a reverberant classroom environment. These aim to simulate typical application use cases, i.e., a hands-free communication and a teleconferencing scenario. For both scenarios, the calculation of the energy decay plots show that the reverberation energy of the resulting overall transfer function is indeed reduced compared with the acoustic transfer functions of the individual channels. Also the direct-to-reverberation ratio is improved for the car as well as the classroom. By comparing the overall transfer function of the partial equalization to the delay-and-sum beamformer, both show a similar magnitude response which suggests a partial equalization of the overall transfer function for both reference choices. In the car environment the G-MWF was simulated in an adaptive manner, which results in a time-varying MWF filter function. Therefore signal-based performance measures for the SNR improvement as well as the dereverberation capabilities are applied in terms of the signal-to-reverberation ratio and the segmental SNR. Both delay-and-sum references show a superior performance regarding the SRR compared with the S-MWF and the P-MWF and are also able to further improve the segmental SNR.

For the differential beamforming references simulations were performed by a monaural hearing aid consisting of two closely spaced microphones. These were used in an anechoic simulation environment as well as a cafeteria. The former allows to create a cylindrically isotropic noise field while the latter contains recordings of real background noise. Simulations results show that the directional beam pattern references for the G-MWF perform superior to the S-MWF in terms of SNR improvement for the anechoic environment and the cafeteria. The highest SSNR gains for both cases are obtained by the hypercardioid reference. The log-spectral distance is preserved for all differential beamforming references compared with the S-MWF. The P-MWF and the DSB reference are also compared to the S-MWF and the simulation results show that no significant SSNR improvement can be achieved, since microphone diversity cannot be exploited due to the closely spaced microphones. By performing angle dependent signal source simulations, the angle dependent SSNR shows that noise sources coming from an incident angle of the spatial null of the dedicated directional beam pattern of the reference are suppressed best. By varying the incident angle of the desired speech source in a cylindrically isotropic noise field, the SSNR improvement is still superior compared with the S-MWF for incident angles that are in the facial field of the person wearing the hearing aid.

The obtained results show, that these new reference choices are able to improve the signal quality further compared with the S-MWF by exploiting the information about the TDOA. By using spatially distributed microphones, the delay-and-sum references benefit from the signal diversity of the sound field to reduce reverberation besides SNR improvement, while the closely spaced microphones of the hearing aid allow to form directional beam patterns to improve the signal quality.

5 Reference for the Binaural Multichannel Wiener Filter

In the last chapter, new reference designs for the multichannel Wiener filter were examined for closely spaced microphones of a monarual hearing aid. This raises the question if these directivity based MWF references can be applied in the context of binaural hearing aids. Research on binaural noise reduction exists for some time [69, 70]. Some known techniques like the MVDR beamformer have been expanded for this use case [71]. With the introduction of the binaural multichannel Wiener filter (BMWF), the concept of multichannel Wiener filtering was adopted for the application in binaural hearing aids [72, 73, 74, 75, 76, 77, 78, 79, 80].

The BMWF and other binaural noise reduction approaches face the challenge of preserving the binaural cues between the left and the right ear. These are described by the interaural transfer functions (ITFs), which allow directional hearing to locate signal sources in the spherical sound field of a human being. The binaural cues consist of the interaural-level-difference (ILD) and the interaural-time difference (ITD) between the two ears [81]. The ILD is caused by the head shadow effect, which leads to a level attenuation at higher frequencies for the ear that is at the far side of a sound source. The ITD is caused due to the time difference of arrival of the signal at the two ears. In [82], it is shown that the ITD is important for the source localization in the lower frequency range, whereas the ILD is important in the higher frequency range.

It has been shown in [74] that the binaural SDW-MWF is able to preserve the binaural cues of a desired speech source. However, the binaural cues of the noise field are distorted. Several efforts have been made to restore the cues of the noise signals. In [73], a BMWF approach with partial noise estimation was presented, which allows for a trade-off between noise reduction and binaural cue preservation of the noise field. This is achieved by mixing the microphone input signals of the left and right reference channel with the output of the BMWF. In [74], the MWF cost function was extended by the ITFs of the speech and noise components to maintain the binaural cues of both

S. Grimm, *Directivity Based Multichannel Audio Signal Processing For Microphones in Noisy Acoustic Environments*, Schriftenreihe der Institute für Systemdynamik (IDS) und optische Systeme (ISO), [illegible]

signals. This concept was further improved by [76], where the average noise ITFs are replaced by the instantaneous ITFs to achieve an improved binaural cue preservation of the speech and noise terms. The work in [77, 78] accounts for isotropic noise fields, which are not considered by ITFs. Therefore an approach was presented to maintain the interaural coherence function of the noise field. Directional interfering noise reduction for the BMWF is presented in [83]. Therefore, the cost function of the BMWF is extended by a constraint to cancel the interfering noise source while preserving the desired signal source. The resulting BMWF equals a subtraction of the interfering components from the standard BMWF implementation.

In this chapter, the directional references for the MWF, which were examined in the previous chapter of this thesis, are applied to binaural hearing aids. Since many efforts in research are made regarding the preservation of the binaural cues, the influence of the direcitvity based references for the MWF on the spatial hearing of a human being is examined. In contrast to [83], directional interference reduction is considered by applying directional references to the BMWF. Since many of the described cue preservation techniques rely on the selection of a reference channel, the directivity based references can theoretically be applied to many of the existing approaches. In this work, the influence of the reference design is investigated for the standard BMWF, which only preserves the binaural cues of a desired speech source.

The chapter is outlined as follows. In section 5.1, the binaural multichannel Wiener filter is described, followed by the description of the binaural cues and the corresponding error criteria in section 5.2. In section 5.3, the generalization of the BMWF is introduced to apply the directional references in context of binaural hearing. Also the influence of the reference design on the binaural cues is derived. In section 5.4, the directional references are applied in two simulation scenarios for binaural hearing aids with two closely spaced microphones at each ear. One simulation environment is anechoic, while the other is a reverberant cafeteria to examine the performance in different acoustic spaces. The obtained results regarding the SSNR improvement and the influence of the directivity based references on the binaural cue preservation are presented. These are followed by a summary in section 5.5.

5.1 The Binaural Multichannel Wiener Filter

The binaural multichannel Wiener filter is an extension of the SDW-MWF. Therefore two filters, $\mathbf{G}_\mathbf{L}^\mathbf{MWF}$ and $\mathbf{G}_\mathbf{R}^\mathbf{MWF}$ are applied to the microphone signals to obtain the output signals for the left and the right hearing aid. For a single desired signal source, the BMWF is defined as

$$\mathbf{G}_\mathbf{L}^\mathbf{MWF} = (\boldsymbol{R}_S + \mu \boldsymbol{R}_N)^{-1} \boldsymbol{R}_S \mathbf{u}_L \tag{5.1}$$

$$\mathbf{G}_\mathbf{R}^\mathbf{MWF} = (\boldsymbol{R}_S + \mu \boldsymbol{R}_N)^{-1} \boldsymbol{R}_S \mathbf{u}_R \tag{5.2}$$

where $\mathbf{u}_L$ and $\mathbf{u}_R$ are the reference vectors for the left and right hearing aid. Analogous to the S-MWF, the standard binaural multichannel Wiener filter (S-BMWF) is defined by setting a one in the reference vector at the corresponding position, while all other entries are set to zero

$$\mathbf{u}_L = [0, \ldots, 1, \ldots, 0]^T \tag{5.3}$$

$$\mathbf{u}_R = [0, \ldots, 1, \ldots, 0]^T . \tag{5.4}$$

The position of the references ref_L and ref_R for the left and the right ear are chosen from the microphone channels as

$$ref_L \in \left[1, \frac{M}{2}\right] \tag{5.5}$$

$$ref_R \in \left[\frac{M}{2} + 1, M\right] , \tag{5.6}$$

where M is an even integer. Often these are chosen pairwise by $ref_R = ref_L + \frac{M}{2}$. Finally, the filtered output signals for the left and the right ear are obtained by

$$Z_L = \mathbf{G}_\mathbf{L}^{\mathbf{MWF}\dagger} \mathbf{Y} \tag{5.7}$$

$$Z_R = \mathbf{G}_\mathbf{R}^{\mathbf{MWF}\dagger} \mathbf{Y} . \tag{5.8}$$

The corresponding system structure of the complete signal processing algorithm is depicted in Figure 5.1.

For the evaluation of the binaural cues, as will be derived in the next section,

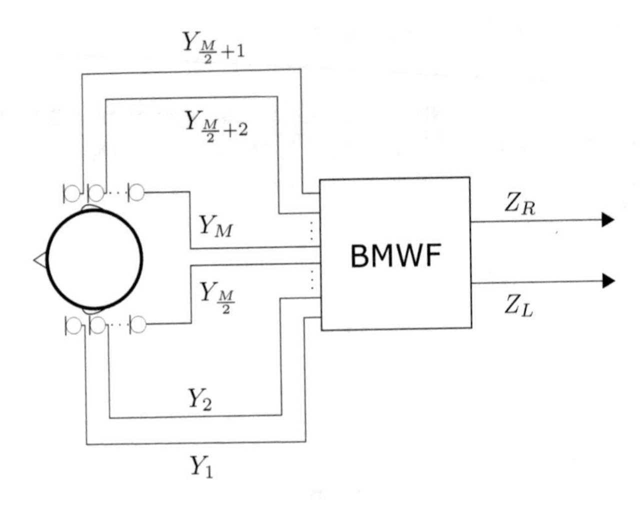

Figure 5.1: System structure of the binaural multichannel Wiener filter

the microphone input signals of the corresponding reference channels are required, which can be obtained by

$$Y_L = \mathbf{u}_L^\dagger \mathbf{Y} = \mathbf{u}_L^\dagger \mathbf{S} + \mathbf{u}_L^\dagger \mathbf{N} \tag{5.9}$$

$$Y_R = \mathbf{u}_R^\dagger \mathbf{Y} = \mathbf{u}_R^\dagger \mathbf{S} + \mathbf{u}_R^\dagger \mathbf{N}\,. \tag{5.10}$$

5.2 Binaural Cues

The binaural cues define the magnitude and phase relationship between the left and right ear for a signal source in the spatial sound field. They are important for a human being to localize one or more signal sources. In the context of hearing aids, the binaural cues are closely related to the interaural transfer functions for two reference channels of the left and right hearing aid. The following definitions and error measures are based on [75].

5.2.1 Interaural Transfer Function

The interaural transfer function (ITF) is the relative transfer function between the left and the right reference channel of a binaural hearing aid. It is distinguished between the input and output ITFs of the processing algorithm. The input ITF of the (single) speech signal is defined as

$$ITF_{\mathbf{S}}^{in} = \frac{\mathbf{u}_L^\dagger \mathbf{S}}{\mathbf{u}_R^\dagger \mathbf{S}} = \frac{\mathbf{u}_L^\dagger \mathbf{H}}{\mathbf{u}_R^\dagger \mathbf{H}} \tag{5.11}$$

$$\tag{5.12}$$

and the output ITF equals

$$ITF_{\mathbf{S}}^{out} = \frac{Z_{SL}}{Z_{SR}} = \frac{\mathbf{G}_{\mathbf{L}}^{\mathbf{MWF}\dagger} \mathbf{S}}{\mathbf{G}_{\mathbf{R}}^{\mathbf{MWF}\dagger} \mathbf{S}} = \frac{\mathbf{G}_{\mathbf{L}}^{\mathbf{MWF}\dagger} \mathbf{H}}{\mathbf{G}_{\mathbf{R}}^{\mathbf{MWF}\dagger} \mathbf{H}}, \tag{5.13}$$

where Z_{SL} and Z_{SR} denote the speech component of the algorithm output for the left and the right ear. The ITFs of the speech signal can also be defined by the correlation matrices

$$ITF_{\mathbf{S}}^{in} = \frac{\mathbf{u}_L^\dagger \boldsymbol{R}_S \mathbf{u}_R}{\mathbf{u}_R^\dagger \boldsymbol{R}_S \mathbf{u}_R} = \frac{\mathbf{u}_L^\dagger \boldsymbol{R}_S \mathbf{u}_L}{\mathbf{u}_R^\dagger \boldsymbol{R}_S \mathbf{u}_L} \tag{5.14}$$

$$ITF_{\mathbf{S}}^{out} = \frac{\mathbf{G}_{\mathbf{L}}^{\mathbf{MWF}\dagger} \boldsymbol{R}_S \mathbf{G}_{\mathbf{R}}^{\mathbf{MWF}}}{\mathbf{G}_{\mathbf{R}}^{\mathbf{MWF}\dagger} \boldsymbol{R}_S \mathbf{G}_{\mathbf{R}}^{\mathbf{MWF}}} = \frac{\mathbf{G}_{\mathbf{L}}^{\mathbf{MWF}\dagger} \boldsymbol{R}_S \mathbf{G}_{\mathbf{L}}^{\mathbf{MWF}}}{\mathbf{G}_{\mathbf{R}}^{\mathbf{MWF}\dagger} \boldsymbol{R}_S \mathbf{G}_{\mathbf{L}}^{\mathbf{MWF}}}, \tag{5.15}$$

and analogous the ITFs of the noise signals $ITF_{\mathbf{N}}^{in}$ and $ITF_{\mathbf{N}}^{out}$ can be defined by replacing the speech correlation matrix $\boldsymbol{R}_S$ in (5.14) and (5.15) with the noise correlation matrix $\boldsymbol{R}_N$.

5.2.2 Binaural Error Measures

To measure derivations and distortions of the binaural cues due to the processing algorithm, the ITFs are separated in magnitude and phase response in terms of the interaural level difference (ILD) and the interaural time difference (ITD). The ILDs for the speech and noise signals are defined as (again, cf. [75])

$$ILD_{\mathbf{S}}^{in} = \frac{\mathbf{u}_L^{\dagger} \boldsymbol{R}_S \mathbf{u}_L}{\mathbf{u}_R^{\dagger} \boldsymbol{R}_S \mathbf{u}_R} \quad (5.16)$$

$$ILD_{\mathbf{N}}^{in} = \frac{\mathbf{u}_L^{\dagger} \boldsymbol{R}_N \mathbf{u}_L}{\mathbf{u}_R^{\dagger} \boldsymbol{R}_N \mathbf{u}_R} \quad (5.17)$$

$$ILD_{\mathbf{S}}^{out} = \frac{\mathbf{G}_{\mathbf{L}}^{\mathbf{MWF}\dagger} \boldsymbol{R}_S \mathbf{G}_{\mathbf{L}}^{\mathbf{MWF}}}{\mathbf{G}_{\mathbf{R}}^{\mathbf{MWF}\dagger} \boldsymbol{R}_S \mathbf{G}_{\mathbf{R}}^{\mathbf{MWF}}} \quad (5.18)$$

$$ILD_{\mathbf{N}}^{out} = \frac{\mathbf{G}_{\mathbf{L}}^{\mathbf{MWF}\dagger} \boldsymbol{R}_N \mathbf{G}_{\mathbf{L}}^{\mathbf{MWF}}}{\mathbf{G}_{\mathbf{R}}^{\mathbf{MWF}\dagger} \boldsymbol{R}_N \mathbf{G}_{\mathbf{R}}^{\mathbf{MWF}}} \quad (5.19)$$

and the ITD as

$$ITD_{\mathbf{S}}^{in} = \angle\left(\frac{\mathbf{u}_L^{\dagger} \boldsymbol{R}_S \mathbf{u}_R}{\mathbf{u}_R^{\dagger} \boldsymbol{R}_S \mathbf{u}_R}\right) = \angle\left(\mathbf{u}_L^{\dagger} \boldsymbol{R}_S \mathbf{u}_R\right) \quad (5.20)$$

$$ITD_{\mathbf{N}}^{in} = \angle\left(\mathbf{u}_L^{\dagger} \boldsymbol{R}_N \mathbf{u}_R\right) \quad (5.21)$$

$$ITD_{\mathbf{S}}^{out} = \angle\left(\mathbf{G}_{\mathbf{L}}^{\mathbf{MWF}\dagger} \boldsymbol{R}_S \mathbf{G}_{\mathbf{R}}^{\mathbf{MWF}}\right) \quad (5.22)$$

$$ITD_{\mathbf{N}}^{out} = \angle\left(\mathbf{G}_{\mathbf{L}}^{\mathbf{MWF}\dagger} \boldsymbol{R}_N \mathbf{G}_{\mathbf{R}}^{\mathbf{MWF}}\right), \quad (5.23)$$

where $\angle(\cdot)$ denotes the phase argument. Based on the ILD and ITD, the corresponding error measures can be derived. The ILD error is calculated for a block length of L samples as

$$\Delta ILD_{\mathbf{S}} = \frac{1}{L}\sum_{\nu=0}^{L-1} 10log_{10}\left(ILD_{\mathbf{S}}^{out}(\nu)\right) - 10log_{10}\left(ILD_{\mathbf{S}}^{in}(\nu)\right) \quad (5.24)$$

$$\Delta ILD_{\mathbf{N}} = \frac{1}{L}\sum_{\nu=0}^{L-1} 10log_{10}\left(ILD_{\mathbf{N}}^{out}(\nu)\right) - 10log_{10}\left(ILD_{\mathbf{N}}^{in}(\nu)\right) \quad (5.25)$$

An ILD value greater than zero indicates a distortion of the ILD. In contrast to the definition in [75], the factor $\frac{1}{L}$ is used to average the ILD over all frequency bins.

The ITD error is defined as

$$\Delta ITD_{\mathbf{S}} = \sum_{\nu=0}^{L-1} \frac{|ITD_{\mathbf{S}}^{out}(\nu) - ITD_{\mathbf{S}}^{in}(\nu)|}{\pi} \tag{5.26}$$

$$\Delta ITD_{\mathbf{N}} = \sum_{\nu=0}^{L-1} \frac{|ITD_{\mathbf{N}}^{out}(\nu) - ITD_{\mathbf{N}}^{in}(\nu)|}{\pi}. \tag{5.27}$$

Note that the ITD error is divided by π to obtain a normalized error value between zero and one.

5.3 The Generalized Binaural Multichannel Wiener Filter

In the following section, the G-MWF, as introduced in chapter 3, is extended for the binaural case. By this generalization, it is possible to implement the reference choices as presented in chapter 4 to obtain an improved noise reduction compared with the S-BMWF. However, the impact on the binaural cues for the combination of the individual channels has to be investigated. Therefore the generalized binaural multichannel Wiener filter (G-BMWF) and its influence on the ITF preservation is derived in the following.

5.3.1 The Generalization of the Binaural Multichannel Wiener Filter

Again, the B-MWF, as defined in (5.2), is used to create the binaural output signals. However, compared with the S-BMWF, the reference vectors $\mathbf{u}_L$ and $\mathbf{u}_R$ not only consist of a single one for the dedicated reference channel but are now extended to a combination of several channels to form the left and right ear output signals. To maintain the binaural cues, one requirement is that only channels of the dedicated ear can be selected, i.e., setting the elements of the vector u_{iL} to zero for $i > \frac{M}{2}$ and the elements of u_{iR} to zero for $i \leq \frac{M}{2}$.

Similar to the monaural case in (3.3), the overall transfer functions for the left and right ear ($\tilde{H}_{dL}$ and $\tilde{H}_{dR}$) can be written as

$$\tilde{H}_{dL} = \mathbf{u}_L^{\dagger}\mathbf{H} = \sum_{i=1}^{\frac{M}{2}} u_{iL}{}^* \cdot H_i \text{ for } u_{iL} \in \mathbb{C} \tag{5.28}$$

$$\tilde{H}_{dR} = \mathbf{u}_R^{\dagger}\mathbf{H} = \sum_{i=\frac{M}{2}+1}^{M} u_{iR}{}^* \cdot H_i \text{ for } u_{iR} \in \mathbb{C}\,. \tag{5.29}$$

5.3.2 On the ITF Preservation of the G-BMWF

Now, the influence of the reference vectors $\mathbf{u}_L$ and $\mathbf{u}_R$ on the binaural cues is derived. The interaural transfer function for a single speech source can be written as

$$ITF_{\mathbf{S}}^{in} = \frac{\mathbf{u}_L^{\dagger}\mathbf{H}}{\mathbf{u}_R^{\dagger}\mathbf{H}}\,, \tag{5.30}$$

which for the S-BMWF is determined by the selection of the dedicated reference channels, i.e.,

$$ITF_{\mathbf{S}}^{in} = \frac{H_{ref_L}}{H_{ref_R}}\,. \tag{5.31}$$

The S-BMWF realization is able to preserve the binaural cues of the speech signal perfectly as has been shown in [74]. The ITFs are now assumed to be identical for each microphone pair, i.e., the ratio of the transfer functions of the left and the right ear are assumed to be pairwise identical for closely spaced microphones

$$ITF_{\mathbf{S}}^{in} = \frac{H_l}{H_{(l+\frac{M}{2})}} \text{ for } l \in \left[1, \frac{M}{2}\right]\,. \tag{5.32}$$

The output of the G-BMWF is a combination of the individual channels for the dedicated ear as stated in (5.28) and (5.29). Therefore, for the G-BMWF the following statement must hold true to preserve the binaural cues of the speech source

$$\frac{\sum_{i=1}^{\frac{M}{2}} u_{iL} H_i}{\sum_{i=1}^{\frac{M}{2}} u_{(i+\frac{M}{2})R} H_{(i+\frac{M}{2})}} \overset{!}{=} \frac{H_l}{H_{(l+\frac{M}{2})}}\,. \tag{5.33}$$

This can be written as

$$\sum_{i=1}^{\frac{M}{2}} u_{iL} H_i = \sum_{i=1}^{\frac{M}{2}} \frac{H_l}{H_{(l+\frac{M}{2})}} u_{(i+\frac{M}{2})R} H_{(i+\frac{M}{2})} \, . \tag{5.34}$$

Due to the assumption that the ratio of the acoustic transfer functions is pairwise identical, as stated in (5.32), this also holds for $l = i$ in (5.34), which results in

$$\sum_{i=1}^{\frac{M}{2}} u_{iL} H_i = \sum_{i=1}^{\frac{M}{2}} u_{(i+\frac{M}{2})R} H_i \, . \tag{5.35}$$

It follows that (5.33) holds if $u_{iL} = u_{(i+\frac{M}{2})R}$ is fulfilled.

This derivation shows that the G-BMWF is indeed able to preserve the binaural cues under the assumption that the ratio of the acoustic transfer functions is pairwise identical. Due to the closely spaced microphones of a hearing aid, this assumption seems valid, since the acoustic transfer functions may mainly differ by a different time-of-arrival of the speech signal. It is shown that it is possible to design the overall transfer functions $\tilde{H}_{dL}$ and $\tilde{H}_{dR}$ by combining the individual channel for the left and right ear. However, to preserve the binaural cues of the speech signal, the combining must be performed in parallel, i.e., $u_{iL} = u_{(i+\frac{M}{2})R}$.

5.4 Simulation Results

In the following, the simulation results for the G-BMWF reference choices are presented. For the simulations, again the database of multichannel in-ear and behind-the-ear head-related and binaural room impulse responses [68] is used. However, now the binaural case is considered. Therefore two microphones for each ear (named front and middle) are investigated ($M = 4$ in total). The sampling rate for all simulations is 48 kHz. A FFT length of $L = 512$ with a blockshift of 128 samples is used with an overlapp-add implementation with a Hamming window. The spacing between the microphones at each hearing aid (left and right ear) is 7.6 mm. Similar as in section 4.6.2, two simulation environments are investigated, namely an anechoic room with a cylindrical diffuse noise field and a cafeteria with babble noise. This allows to examine the acoustic influence on the binaural cues. The PSDs and CPSDs as well as the TDOA are assumed to be known and the G-BMWF is applied as a batch job. An overestimation parameter $\mu = 1$ is used for all simulations and the desired speech source is always coming from an incident angle of $\theta = 0°$.

The reference vectors for the G-BMWF, as used in the simulations, are presented in Table 5.1. This aims to make the naming more clear for the discussion of the simulation results in the following.

Table 5.1: G-BMWF reference realization for $M = 4$ microphones

Name of Reference choice	Realization of $\mathbf{u}$
S-BMWF (left)	$\mathbf{u}_L = [1, 0, 0, 0]^T$
S-BMWF (right)	$\mathbf{u}_R = [0, 0, 1, 0]^T$
G-BMWF-DS (left)	$\mathbf{u}_L = [e^{j2\pi\nu\tau_0}, 1, 0, 0]^T$
G-BMWF-DS (right)	$\mathbf{u}_R = [0, 0, e^{j2\pi\nu\tau_0}, 1]^T$
P-BMWF (left)	$\mathbf{u}_L = \left[\sqrt{\frac{r_{S_{1,1}}}{(r_{S_{1,1}}+r_{S_{2,2}})}}, \sqrt{\frac{r_{S_{2,2}}}{(r_{S_{1,1}}+r_{S_{2,2}})}}\frac{r_{S_{2,1}}}{\lvert r_{S_{2,1}}\rvert}, 0, 0\right]^T$
P-BMWF (right)	$\mathbf{u}_R = \left[0, 0, \sqrt{\frac{r_{S_{3,3}}}{(r_{S_{3,3}}+r_{S_{4,4}})}}, \sqrt{\frac{r_{S_{4,4}}}{(r_{S_{3,3}}+r_{S_{4,4}})}}\frac{r_{S_{4,3}}}{\lvert r_{S_{4,3}}\rvert}\right]^T$
BMWF-Dipole reference (left)	$\mathbf{u}_L = G^{EQ^*}[1, -1, 0, 0]^T$
BMWF-Dipole reference (right)	$\mathbf{u}_R = G^{EQ^*}[0, 0, 1, -1]^T$
BMWF-Cardioid reference (left)	$\mathbf{u}_L = G^{EQ^*}[1, -e^{j2\pi\nu\tau_0}, 0, 0]^T$
BMWF-Cardioid reference (right)	$\mathbf{u}_R = G^{EQ^*}[0, 0, 1, -e^{j2\pi\nu\tau_0}]^T$
BMWF-Hypercardioid reference (left)	$\mathbf{u}_L = G^{EQ^*}[1, -e^{j2\pi\nu\frac{\tau_0}{3}}, 0, 0]^T$
BMWF-Hypercardioid reference (right)	$\mathbf{u}_R = G^{EQ^*}[0, 0, 1, -e^{j2\pi\nu\frac{\tau_0}{3}}]^T$

5.4.1 SSNR and LSD

In Table 5.2, the results for the segmental SNR, as calculated in (4.25), are presented for several reference choices of the G-BMWF. Similar to the results in chapter 4, the SSNR is calculated for frequencies between 400 Hz and 4 kHz by band-limiting the signals. For each reference choice, the SSNR values for the outputs of the left and right hearing aid are shown. For the S-BMWF reference selection, the front microphones of the left and right hearing aid are chosen (Y_1 and Y_3). The P-BMWF uses only the phase term of Y_1 and Y_3 as the reference due to the magnitude combining. The results show the SSNR values for the anechoic as well as the cafeteria environment. It should be noted that the achieved noise reduction is higher than the results in Table 4.3, since the number of microphones used by the MWF is doubled for the binaural case. As can be observed, the SSNR is improved for the differential beamforming references in comparison with the S-BMWF in both simulation scenarios. The hypercardioid reference achieves the best SSNR values for both simulation environments. The G-BMWF-DS, that uses a delay-and-sum reference, as well as the P-BMWF show only a small improvement compared with the S-BMWF, since the microphones are closely spaced.

In Table 5.3, the LSD values for the different reference choices are presented. The results show similar behavior compared with Table 4.4. The differential beamforming reference choices are able to slightly improve the LSD for the anechoic environment. For the cafeteria the LSD values are similar to the microphone input signals. Only the dipole reference shows an increased LSD value. The results indicate that the G-BMWF improves the LSD for some cases but shows at least similar values compared with the S-BMWF.

5.4.2 Binaural Cues

In the following, the influence of the reference choices on the binaural cues is examined. Therefore the ITF of the speech signal for the S-BMWF reference selections as well as for several directivity based G-BMWF references is calculated from the resulting overall transfer functions. Note that all results are shown for a speech source coming from the front ($\theta = 0°$) and are not necessary valid for other incident angles.

ITF comparison

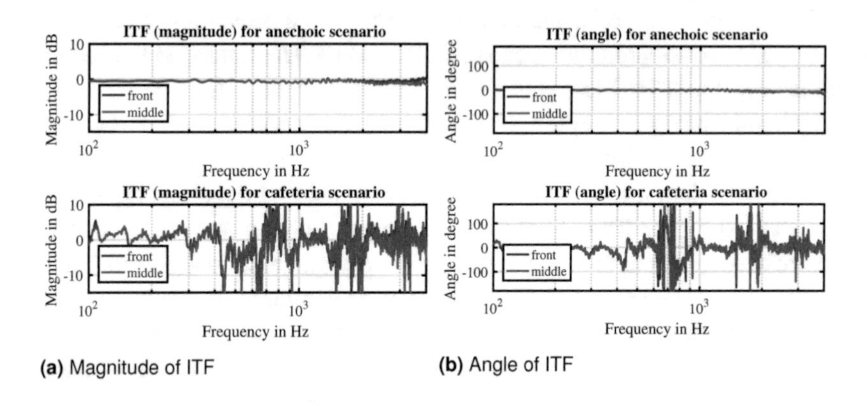

(a) Magnitude of ITF **(b)** Angle of ITF

Figure 5.2: ITF comparison - front and middle microphones

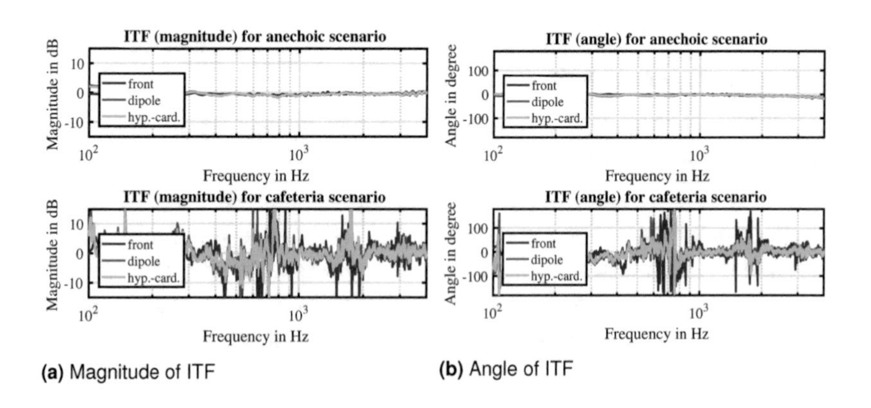

(a) Magnitude of ITF **(b)** Angle of ITF

Figure 5.3: ITF comparison - various references

In Figure 5.2, the ITFs of the speech signal are depicted for the anechoic environment as well as the cafeteria (incident angle of the speech source $\theta = 0°$). As can be observed, the magnitude as well as the phase are nearly identical when the front microphones (Y_1 and Y_3) or the middle microphones (Y_2 and Y_4) are chosen as the ITF references (S-BMWF). Based on these plots, for the anechoic environment as well as for the cafeteria, the assumption about pairwise identical acoustic transfer function ratios seems valid despite some small deviations.

Figure 5.3 shows the resulting ITFs of the speech signal for several directivitiy based reference choices. As can be seen, the magnitude as well as the phase are nearly identical for all reference choices in the anechoic environment. For the cafeteria, however, small deviations of the magnitude as well as the phase can be observed due to the acoustic influences.

ITD and ILD error

Further, the ITD and the ILD errors are calculated for several G-BMWF reference choices. In Table 5.4, the results regarding the ITD errors for the speech as well as for the noise signal are presented for the anechoic environment. As can be seen, the ITD error of the speech signal is close to zero for all G-BMWF references with small error values for the differential references. This shows that the directivity based references are able to improve the SSNR compared with the S-BMWF, while still retaining the binaural cues. For all cases, the ITD error of the noise signal is quite high, since this BMWF implementation distorts the binaural cues of the noise signal as already stated in [74]. However, the directivity based references have no significant impact on the noise ITD error.

For the cafeteria, the results regarding the ITD error are presented in Table 5.5. While the G-BMWF-DS and the P-BMWF implementation have no influence on the ITD error, the differental beamforming references show a slightly increased error value for the speech ITD. By comparison with the anechoic environment, this seems to be caused by the acoustic influences of the room. The noise ITF error, however, is not affected by the reference choice.

In Table 5.6 and Table 5.7 the ILD error for the anechoic environment as well as the cafeteria are shown. As can be observed, the error is slightly increased for the directivity based references for the speech and noise ILD in both scenarios. The dipole reference seems to have the worst performance regarding the ILD error, since for both simulation scenarios the value is slightly increased.

The simulation result show, that the G-BMWF is able to further improve the SSNR as well as the LSD compared with the S-BMWF by using the directivity based reference choices. As has been shown, the binaural cues of the speech signal in terms of the ITF, the ITD and the ILD can be preserved for these references in the anechoic environment. This indicates that the assumption about the pairwise identical transfer function ratio seems valid for this case. For the reverberant cafeteria the ITFs show slight deviations if the directivity based references are used to design the overall transfer functions of the left and the right hearing aid. By comparing the ITD and ILD error for the anechoic environment with the reverberant cafeteria, the acoustic influences seem to slightly decrease the performance. As a general conclusion, the SSNR can be improved in comparison with the S-BMWF for the directivity based references at the expense of a slight distortion of the binaural cues.

5.5 Summary

In this chapter, the multichannel Wiener filter was applied in the context of binaural noise reduction for the application of hearing aids. The processing algorithms face the challenge to preserve the binaural cues of the desired speech source as well as the noise field. As stated in the introduction of this chapter, several efforts on binaural cue preservation have been made in the literature. The S-BMWF therefore uses a reference channel of the left and right ear to preserve the binaural cues of the desired speech source. However, the binaural cues of the noise field are distorted. By extending the S-BMWF to the G-BMWF, the reference choice for the left and right ear allows a combining of the individual channels to design the dedicated overall transfer function. Under the assumption that the acoustic transfer function ratios of the microphone signals are pairwise identical, the combined reference choices have been investigated regarding their influence on the binaural cues. As has been derived, the binaural cues can be preserved if the reference vectors are applied in parallel for the left and the right ear. This allows to use the directivity based references of the G-MWF in the context of binaural noise reduction.

To verify these derivations, simulations were performed for a binaural hearing aid consisting of two closely spaced microphones for each ear. An anechoic environment as well as a reverberant cafeteria are considered to examine the acoustic influences on the performance. It was shown that the directivity based references which use differential beamforming are able to improve the SSNR and for some cases the LSD in both simulation environments compared with the S-BMWF.

By comparing the interaural transfer functions, it can be observed that these are pairwise identical for the dedicated microphone channels which makes this assumption valid for the examined simulation scenarios. Further, the speech signal ITFs are compared to the ITFs of the G-BMWF overall transfer functions that use the directivity based references. It can be shown that these are identical for the anechoic environment. However, slight deviations for the more reverberant cafeteria environment can be observed. The results of the ITD and the ILD error measures show that the S-BMWF preserves the binaural cues of the desired speech signal perfectly while distorting the binaural cues of the noise signal as already stated in the literature. However, the errors for the G-BMWF using the directivity based reference selections are nearly identical compared with the S-BMWF with a small increase for the binaural cue errors of the speech signal in the reverberant cafeteria environment.

This leads to the conclusion that the directivity based reference choices can be applied in the context of binaural noise reduction while preserving the binaural cues of a desired speech signal as long as these references are applied in parallel. For more reverberant environments the assumption about pairwise identical interaural transfer functions is slightly violated which results in a slight distortion of the binaural cues, however, the noise reduction capabilities are superior compared with the S-BMWF.

Table 5.2: Segmental SNR for different G-BMWF references

$\mu = 1$	Anechoic environment	Cafeteria environment
Y_1 (left - front)	5.6 dB	4.2 dB
Y_2 (left - middle)	5.5 dB	4.1 dB
Y_3 (right - front)	5.7 dB	4.3 dB
Y_4 (right - middle)	5.8 dB	4.4 dB
S-BMWF (left)	8.0 dB	5.8 dB
S-BMWF (right)	8.2 dB	5.8 dB
G-BMWF-DS (left)	8.1 dB	5.7 dB
G-BMWF-DS (right)	8.5 dB	5.8 dB
P-BMWF (left)	8.2 dB	5.7 dB
P-BMWF (right)	8.4 dB	5.7 dB
BMWF-Dipole reference (left)	8.9 dB	7.2 dB
BMWF-Dipole reference (right)	9.0 dB	7.3 dB
BMWF-Cardioid reference (left)	9.2 dB	7.3 dB
BMWF-Cardioid reference (right)	9.1 dB	7.4 dB
BMWF-Hypercardioid reference (left)	9.4 dB	7.8 dB
BMWF-Hypercardioid reference (right)	9.5 dB	7.9 dB

Table 5.3: Log spectral distance for different G-BMWF references

$\mu = 1$	Anechoic environment	Cafeteria environment
Y_1 (left - front)	3.2 dB	3.8 dB
Y_2 (left - middle)	2.7 dB	3.6 dB
Y_3 (right - front)	3.1 dB	3.6 dB
Y_4 (right - middle)	2.8 dB	3.5 dB
S-BMWF (left)	2.7 dB	4.3 dB
S-BMWF (right)	2.7 dB	4.1 dB
G-BMWF-DS (left)	2.6 dB	4.2 dB
G-BMWF-DS (right)	2.6 dB	4.0 dB
P-BMWF (left)	2.6 dB	4.2 dB
P-BMWF (right)	2.5 dB	4.1 dB
BMWF-Dipole reference (left)	1.9 dB	4.0 dB
BMWF-Dipole reference (right)	2.0 dB	4.8 dB
BMWF-Cardioid reference (left)	2.2 dB	3.9 dB
BMWF-Cardioid reference (right)	2.3 dB	3.9 dB
BMWF-Hypercardioid reference (left)	2.1 dB	3.9 dB
BMWF-Hypercardioid reference (right)	2.1 dB	4.3 dB

Table 5.4: ITD error for different G-BMWF references - anechoic environment

$\mu = 1$	ΔITD - Speech	ΔITD - Noise
S-BMWF	0 %	45 %
G-BMWF-DS	0 %	45 %
P-BMWF	0 %	49 %
BMWF-Dipole reference	3 %	46 %
BMWF-Cardioid reference	2 %	47 %
BMWF-Hypcard. ref.	3 %	46 %

Table 5.5: ITD error for different G-BMWF references - cafeteria environment

$\mu = 1$	ΔITD - Speech	ΔITD - Noise
S-BMWF	1 %	20 %
G-BMWF-DS	1 %	20 %
P-BMWF	1 %	30 %
BMWF-Dipole reference	9 %	19 %
BMWF-Cardioid reference	6 %	24 %
BMWF-Hypcard. ref.	8 %	23 %

Table 5.6: ILD error for different G-BMWF references - anechoic environment

$\mu = 1$	ΔILD - Speech	ΔILD - Noise
S-MWF	0.1 dB	0.1 dB
G-MWF-DS	0.2 dB	0.2 dB
P-MWF	0.2 dB	0.9 dB
Dipole reference	0.6 dB	0.6 dB
Cardioid reference	0.3 dB	0.4 dB
BMWF-Hypcard. ref.	0.4 dB	0.7 dB

Table 5.7: ILD error for different G-BMWF references - cafeteria environment

$\mu = 1$	ΔILD - Speech	ΔILD - Noise
S-MWF	0.2 dB	0.3 dB
G-MWF-DS	0.3 dB	0.4 dB
P-MWF	0.3 dB	1.1 dB
Dipole reference	1.3 dB	1.4 dB
Cardioid reference	0.7 dB	0.5 dB
BMWF-Hypcard. ref.	0.9 dB	1.0 dB

6 Wind Noise Reduction for a Closely Spaced Microphone Array

In the previous chapters, directivity based references for the generalized multichannel Wiener filter have been discussed. These include the differential beamforming references, which were successfully applied in the context of closely spaced microphone arrangements for hearing aids. However, differential microphone arrays (and therefore the differential references for the G-MWF) are not ideal regarding noise reduction in the presence of wind. Wind noise can for example occur in hands-free communication applications in a car environment and are caused by open windows, fans or open convertible hoods. Also hearing aids can suffer from wind noise if they are worn outdoors. The induced wind creates airflow turbulence over the microphone membranes which results in low frequency signal components of high amplitude [84].

As has been shown in the literature, the wind noise terms are decorrelated between the microphones for distances of even a few centimeters [85, 86]. These correlation properties may lead to a significant amplification of the wind noise for differential beamforming [25]. The required first order low pass filter for the equalization regarding the speech signal makes this behavior even worse. The proposed G-MWF references for closely spaced microphones in hearing aids, as introduced in the previous chapters, are therefore not suitable at all for wind noise reduction. One proposed solution for a differential microphone array is to switch to a single microphone with an omnidirectional response if wind noise is detected to avoid the wind noise amplification [25].

Commonly used noise reduction algorithms are typically based on the assumption that the noise is stationary or varies only slowly in time. In [87], Wilson et. al. demonstrated that wind noise consists of local short time disturbances which are highly non-stationary. This makes the reduction of wind noise a challenging task due to the estimation of the short time PSDs. The suppression of wind noise is mostly covered in the context of digital hearing aids or mobile

S. Grimm, *Directivity Based Multichannel Audio Signal Processing For Microphones in Noisy Acoustic Environments*, Schriftenreihe der Institute für Systemdynamik (IDS) und optische Systeme (ISO), https://doi.org/10.1007/978-3-658-25152-9_6

devices in the literature [25, 88, 89]. For single channel wind noise reduction, often the different power spectral density properties of speech and wind noise are exploited [25, 88, 90]. Several other methods exist that aim to reduce wind noise for a single microphone [91, 92, 93, 94, 95].

However, the utilization of more than one microphone allows to take the diversity of the sound field into account to indicate wind noise and reduce it successfully. In [88], a spectral weighting filter based on the coherence between two microphones is proposed. The coherence is also used in [86], where additionally to the magnitude squared coherence (MSC) the information that relies on the phase component is applied to synthesize a spectral filter function.

In [96], the decomposition of the multichannel Wiener filter into a minimum variance distortionless response beamformer and a single channel Wiener post filter for an arbitrary microphone arrangement is presented. The approach is based on the assumption that the wind noise is decorrelated at the microphones, while having equal noise power spectral densities but arbitrary acoustic transfer functions for the speech signal. From these assumptions follows for closely spaced microphones that a simple delay-and-sum beamformer achieves maximum SNR beamforming, because equal ATFs from the speech source to the microphones can be assumed for low frequencies.

In this chapter, a wind noise reduction approach for a closely spaced microphone array is proposed. The decomposition of the MWF in a beamformer and a single channel post filter is used similar to [96]. It is also assumed that the wind noise terms are decorrelated at the microphones. However, in contrast to [96], it is assumed that the short time noise PSDs at the microphones may differ. If the geometry of the microphone array as well as the location of the desired speech source is known, assumptions about the speech and noise signal properties can be made to design a low-complexity wind noise reduction algorithm.

The use of micro-electro-mechanical system (MEMS) microphones as a replacement for ordinary microphone capsules have gained interest in [97, 98, 99], especially for the application of directive beamforming [100, 101] due to its reduced size and cost compared with an ordinary microphone capsule. This makes them an interesting technology for multichannel speech signal processing. Therefore the proposed wind noise reduction approach is examined for an array of two closely spaced MEMS microphones in a car environment.

This chapter is outlined as follows. The proposed wind noise reduction approach is derived in section 6.1 based on the decomposition of the multichannel Wiener filter. Some special cases for the beamformer are discussed based on assumptions about the wind noise signal properties. In section 6.2, the simulation results for the closely spaced microphone array in a car environment are discussed, followed by a summary in section 6.3.

This chapter has been published in [14].

6.1 Wind Noise Reduction Algorithm

In this section, the proposed noise reduction algorithm is derived. The filtering is only applied in the low frequency range which is affected by wind noise. It should be noted, that the noise signal consists of wind as well as car noise components. However, in the presence of wind noise, the wind noise components are dominant at low frequencies. In the following, only the instationary wind noise components at low frequencies are considered and the slowly varying driving noise is neglected. Such stationary noise components can be estimated and reduced by state-of-the-art noise reduction approaches.

As already stated, the $\mathbf{G}^{\mathbf{MWF}}$ can be decomposed into a MVDR beamformer

$$\mathbf{G}^{\mathbf{MVDR}} = \frac{\boldsymbol{R}_N^{-1}\mathbf{H}}{\mathbf{H}^{\dagger}\boldsymbol{R}_N^{-1}\mathbf{H}} \tag{6.1}$$

and a single channel Wiener post filter

$$G^{WF} = \frac{\gamma_{beam}^{out}}{\gamma_{beam}^{out} + \mu} \tag{6.2}$$

as

$$\mathbf{G}^{\mathbf{MWF}} = \mathbf{G}^{\mathbf{MVDR}} \cdot G^{WF} \cdot \tilde{H}_d^*. \tag{6.3}$$

The term γ_{beam}^{out} is the narrow-band SNR at the beamformer output which is defined as

$$\gamma_{beam}^{out} = \mathrm{tr}(\boldsymbol{R}_S\boldsymbol{R}_N^{-1}), \tag{6.4}$$

where $\mathrm{tr}(\cdot)$ denotes the trace operator. This decomposition is exploited for the proposed wind noise reduction. Firstly, a beamformer for the considered microphone setup is derived.

6.1.1 Beamformer

In the following, time aligned signals are considered, where the alignment compensates the different times of arrival for the speech signal. This is achieved by delaying the front microphone with a suitable sample delay τ to be in phase with the rear microphone

$$\hat{Y}_1(\nu) = Y_1(\nu) \cdot \begin{cases} e^{-j2\pi\frac{\nu}{L}\tau} & \text{for } \nu \in 0, \ldots, \frac{L}{2}-1 \\ e^{j2\pi\frac{\nu}{L}\tau} & \text{for } \nu \in \frac{L}{2}, \ldots, L-1\,, \end{cases} \quad (6.5)$$

where L denotes the block length of the short time Fourier transform. The aligned acoustic transfer function $\hat{H}_1$ is described similarly. After the time alignment, the ATFs in $\mathbf{H}$ are assumed to be identical, because the low frequency speech components have a large wavelength compared with the microphone distance

$$H = \hat{H}_1 = H_2 \quad (6.6)$$

$$\mathbf{H} = H \cdot [1,1]^T\,, \quad (6.7)$$

which leads to the speech correlation matrix depending only on the PSD of the speech signal at one of the microphones

$$\boldsymbol{R}_S = \Phi_X^2 |H|^2 \begin{pmatrix} 1 & 1 \\ 1 & 1 \end{pmatrix} = \Phi_S^2 \begin{pmatrix} 1 & 1 \\ 1 & 1 \end{pmatrix}. \quad (6.8)$$

Furthermore, it can be assumed that the wind noise terms for both microphone signals are uncorrelated even for small distances of the microphones as examined in [85, 86]. This simplifies the noise correlation matrix as well as its inverse since the cross-terms can be neglected

$$\boldsymbol{R}_N^{-1} = \begin{pmatrix} \frac{1}{\Phi_{N_1}^2} & 0 \\ 0 & \frac{1}{\Phi_{N_2}^2} \end{pmatrix}. \tag{6.9}$$

The nominator term of the $\mathbf{G}^{\mathbf{MVDR}}$ in (6.1) can be written as

$$\boldsymbol{R}_N^{-1}\mathbf{H} = H \cdot \begin{pmatrix} \frac{1}{\Phi_{N_1}^2} \\ \frac{1}{\Phi_{N_2}^2} \end{pmatrix} \tag{6.10}$$

and the denominator as

$$\mathbf{H}^{\dagger}\boldsymbol{R}_N^{-1}\mathbf{H} = |H|^2 \cdot \left(\frac{1}{\Phi_{N_1}^2} + \frac{1}{\Phi_{N_2}^2} \right). \tag{6.11}$$

Since H is not known, it is set to $H = 1$. This results in the minimum variance (MV) beamformer coefficients

$$G_i^{MV} = \frac{\frac{1}{\Phi_{N_i}^2}}{\frac{1}{\Phi_{N_1}^2} + \frac{1}{\Phi_{N_2}^2}}, \tag{6.12}$$

which can be interpreted as a noise dependent weighting of the input signals. Note that the MV beamformer achieves the same narrow-band output SNR as the MVDR beamformer but no distortion-free response [31]. Finally, the output of the beamformer can be written as

$$Y_{MV} = (\hat{Y}_1 \cdot G_1^{MV} + Y_2 \cdot G_2^{MV}). \tag{6.13}$$

Using (6.8) and (6.9), it follows for the narrow-band output SNR of the beamformer

$$\gamma_{beam}^{out} = \Phi_S^2 \cdot \left(\frac{1}{\Phi_{N_1}^2} + \frac{1}{\Phi_{N_2}^2} \right) = \frac{\Phi_S^2}{\Phi_{N_{beam}}^2}, \tag{6.14}$$

where $\Phi_{N_{beam}}^2$ denotes the noise PSD at the beamformer output. This PSD can be calculated as

$$\Phi_{N_{beam}}^2 = \frac{\Phi_{N_1}^2 \cdot \Phi_{N_2}^2}{\Phi_{N_1}^2 + \Phi_{N_2}^2}. \tag{6.15}$$

6.1.2 Special Cases

In the following, some special cases for the beamformer as derived in (6.13) are considered. Assuming $\Phi^2_{N_1} = \Phi^2_{N_2}$ and uncorrelated noise terms as in [96], G_i^{MV} reduces to the simple weighting of a delay-and-sum beamformer (a simple summing of the aligned signals), which is also proposed in [25] as the optimal combining in case of wind noise for closely spaced microphones

$$G_i^{DS} = \frac{\frac{1}{\Phi^2_{N_1}}}{\frac{1}{\Phi^2_{N_1}} + \frac{1}{\Phi^2_{N_1}}} = \frac{1}{2}, \tag{6.16}$$

which results in the output signal

$$Y_{DS} = \frac{1}{2}(\hat{Y}_1 + Y_2). \tag{6.17}$$

The condition of uncorrelated noise terms is kept and a special case is assumed, where the short time noise PSDs are varying over time and frequency. This is motivated by the highly non-stationary local short time wind noise disturbances as examined in [87], which implies that only one microphone is affected by wind noise for a certain time and frequency index κ and ν

$$\Phi^2_{N_1}(\kappa,\nu) << \Phi^2_{N_2}(\kappa,\nu) \tag{6.18}$$

or

$$\Phi^2_{N_1}(\kappa,\nu) >> \Phi^2_{N_2}(\kappa,\nu). \tag{6.19}$$

Then the noise PSD dependent weighting in (6.12) reduces to a selection approach of the dedicated frequency bins by comparing the short time PSDs of the microphone signals $\Phi^2_{Y_i}$, because the speech signal PSDs $\Phi^2_{S_i}$ are assumed to be identical for both microphones. Therefore the resulting output signal Y_{FBS} can be written as

$$Y_{FBS}(\kappa,\nu) = \begin{cases} Y_1(\kappa,\nu), & \Phi^2_{Y_1}(\kappa,\nu) < \Phi^2_{Y_2}(\kappa,\nu) \\ Y_2(\kappa,\nu), & \Phi^2_{Y_1}(\kappa,\nu) > \Phi^2_{Y_2}(\kappa,\nu). \end{cases} \tag{6.20}$$

6.1.3 PSD estimation

Next, the estimates for the speech and noise PSDs, which are required for the beamformer and post filter, are derived. As mentioned in [96], most single channel noise estimation procedures (i.e. [55, 102, 57]) rely on the assumption that the noise signal PSDs are varying more slowly in time than the speech signal PSD. This is not the case for wind noise and therefore makes it a challenging task for a single microphone. However, the different correlation properties for speech and wind noise using more than one microphone can be used for the varying short-time PSD estimates.

Considering the small distance of the microphones, a reference for the wind noise can be obtained by exploiting the fact that the wind noise components in the two microphones are incoherent, while the speech components are highly coherent. To block the speech signal, a delay-and-subtract approach is used to get the term

$$N = \frac{\hat{Y}_1 - Y_2}{2}, \tag{6.21}$$

which depends only on incoherent wind noise terms.

The wind noise PSD is defined as

$$\begin{aligned}
\Phi_N^2 &= \mathbb{E}\{NN^*\} && (6.22)\\
&= \mathbb{E}\left\{\left(\frac{\hat{Y}_1 - Y_2}{2}\right)\left(\frac{\hat{Y}_1 - Y_2}{2}\right)^*\right\} && (6.23)\\
&= \frac{1}{4}\left(\mathbb{E}\left\{\hat{Y}_1\hat{Y}_1^*\right\} - \mathbb{E}\left\{\hat{Y}_1 Y_2^*\right\} - \mathbb{E}\left\{Y_2\hat{Y}_1^*\right\} + \mathbb{E}\{Y_2 Y_2^*\}\right) && (6.24)\\
&= \frac{1}{4}\Big(\mathbb{E}\left\{\hat{N}_1\hat{N}_1^*\right\} - \mathbb{E}\left\{\hat{N}_1 N_2^*\right\} && (6.25)\\
&\quad -\mathbb{E}\left\{N_2\hat{N}_1^*\right\} + \mathbb{E}\{N_2 N_2^*\}\Big).
\end{aligned}$$

Since the wind noise terms are assumed to be uncorrelated in the low frequency range, the wind noise cross-terms vanish and the PSD is obtained by

$$\Phi_N^2 = \frac{\Phi_{N_1}^2}{4} + \frac{\Phi_{N_2}^2}{4}. \tag{6.26}$$

(6.21) is used as an estimate for the wind noise at the output of the beamformer. Note that this delay-and-subtract signal in combination with an additional low pass compensation filter is also the output signal of the G-MWF for the differential beamforming references proposed in chapter 4 and chapter 5 (despite the different look direction of the array). Obviously, this is not suitable for microphone positions that are sensitive to wind noise, because the noise terms are heavily amplified.

By summing the aligned signals according to (6.17), the coherent signal components are augmented. The combined signal Y_{DS} has the PSD

$$\begin{aligned}
\Phi^2_{Y_{DS}} &= \mathbb{E}\{Y_{DS}Y^*_{DS}\} && (6.27)\\
&= \mathbb{E}\left\{\left(\frac{\hat{Y}_1+Y_2}{2}\right)\left(\frac{\hat{Y}_1+Y_2}{2}\right)^*\right\} && (6.28)\\
&= \frac{1}{4}\Big(\mathbb{E}\left\{\hat{Y}_1\hat{Y}_1^*\right\}+\mathbb{E}\left\{\hat{Y}_1Y_2^*\right\} && (6.29)\\
&\quad +\mathbb{E}\left\{Y_2\hat{Y}_1^*\right\}+\mathbb{E}\left\{Y_2Y_2^*\right\}\Big)\\
&= \mathbb{E}\left\{SS^*\right\}+\frac{1}{4}\Big(\mathbb{E}\left\{\hat{N}_1\hat{N}_1^*\right\}+\mathbb{E}\left\{\hat{N}_1N_2^*\right\}+\mathbb{E}\left\{N_2\hat{N}_1^*\right\}\\
&\quad +\mathbb{E}\left\{N_2N_2^*\right\}\Big). && (6.30)
\end{aligned}$$

Again the wind noise cross-terms are assumed to vanish due to the uncorrelated wind noise behaviour and from (6.30) the PSD is obtained

$$\Phi^2_{Y_{DS}} = \Phi^2_S + \frac{\Phi^2_{N_1}}{4} + \frac{\Phi^2_{N_2}}{4}. \tag{6.31}$$

Combining (6.26) and (6.31), the PSD of the clean speech signal is acquired by

$$\Phi^2_S = \Phi^2_{Y_{DS}} - \Phi^2_N \tag{6.32}$$

and the noise PSD at the ith microphone

$$\Phi^2_{N_i} = \Phi^2_{Y_i} - \Phi^2_S. \tag{6.33}$$

It should be noted that this derivation only holds for uncorrelated noise terms. Φ^2_S may still contain correlated noise. However, the correlated driving noise is neglected as stated at the beginning of this section. In contrast to Zelinskis post-filter [103], which also assumes zero correlation between the microphone signals, the short time noise PSDs are assumed to be different ($\Phi^2_{N_1} \neq \Phi^2_{N_2}$).

6.1.4 Post Filter

As described in (6.3), the beamformer is followed by a single channel Wiener post filter to achieve additional noise suppression. The SNR estimate

$$\gamma^{in} = \frac{\Phi^2_S}{\Phi^2_N} \tag{6.34}$$

is used with the noise PSD according to (6.26) instead of (6.14), because it shows a better performance in the simulations regarding SNR improvement and speech distortion. Hence, the post filter

$$G^{WF} = \frac{\gamma^{in}}{\gamma^{in} + \mu} \tag{6.35}$$

is used.

Finally, the output of the complete wind noise reduction algorithm is

$$\begin{aligned} Z &= (\hat{Y}_1 \cdot G_1^{MV} + Y_2 \cdot G_2^{MV}) \cdot G^{WF} && (6.36) \\ &= Y_{MV} \cdot G^{WF}. && (6.37) \end{aligned}$$

This wind noise reduction algorithm is only applied for frequencies below a cutoff frequency f_c, because wind noise mostly contains low frequency components and the assumptions about the signal properties are only valid for low frequencies. Figure 6.1 shows the block diagram of the signal processing structure.

6.2 Simulation Results

In the following, the results of the simulations for the algorithm proposed in section 6.1 are presented. Therefore wind noise in a car was recorded with a linear

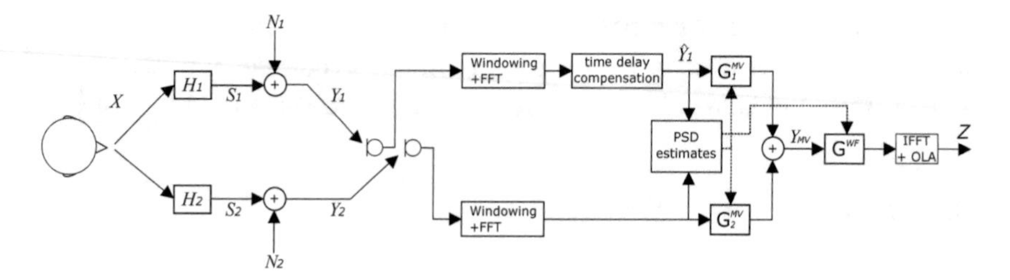

Figure 6.1: Block diagram of the signal model and the proposed processing [14]

MEMS microphone array in endfire configuration. To investigate varying microphone distances, an array of four sensors with equal spacing is considered. The distances are 7.1 mm, 14.3 mm and 21.4 mm. The driving speed was 100 km/h and both front windows at the driver side as well as the co-driver side were completely open to allow a turbulence airflow over the MEMS array. The speech signals for testing are ITU speech signals convolved with the impulse responses, which were measured from the mouth reference point of an artificial head (HMS II.5 from HEAD acoustics) at the drivers position to the MEMS array microphones. The array is mounted above the sun visor at the driver seat position. For the simulations a sampling rate $fs = 16$ kHz and a FFT size of 512 samples is used. The FFT shift is 128 samples and each block is windowed with a Hamming window before it is transformed into the frequency domain. The noise recordings and the speech recordings were done separately and mixed in the simulation. The overestimation parameter for the Wiener post filter is set to $\mu = 6$ and the cutoff frequency f_c is set to 1 kHz.

As quality measures, the segmental signal-to-noise ratio as well as the log spectral distance are considered. It should be noted that all SSNR and LSD measures are calculated for the frequency region below the cutoff frequency f_c since the frequency region above f_c is not affected by the proposed wind noise reduction approach. Car noise, which is also present in the microphone signals, is not considered in the algorithm. Therefore the SSNR improvements in absolute value can be lower compared with measured noise signals which contain wind noise only.

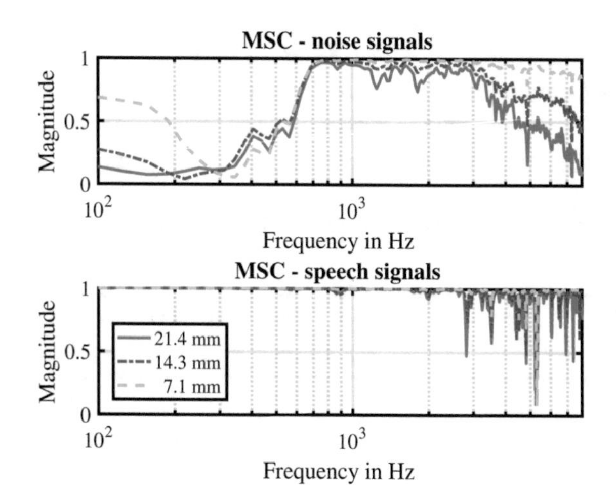

Figure 6.2: Magnitude squared coherence for noise signals (top) as well as speech signals (bottom) with different microphone distances [14]

6.2.1 Coherence Properties

Figure 6.2 shows the results of the MSC calculation of speech and noise for varying microphone distances. As can be observed, the assumption that noise is decorrelated while speech is highly correlated is fulfilled for frequencies below 600 Hz for all microphone distances. Therefore the assumptions that are made for the design of the beamformer as well as for the calculation of the speech and noise estimates are valid for the low frequency range.

6.2.2 Beamformer Output

In Table 6.1 the SSNR gain of the beamformer output is compared with a single microphone. This comparison is considered, because the approach in [25] suggests to switch from a differential microphone array to a single omnidirectional microphone if wind noise is detected. The SSNR of the single microphone is 2.1 dB. For further comparison, the results of the delay-and-sum beamformer Y_{DS} are shown, which is the summing of the aligned signals, as described in (6.17) (and also proposed in [25] for combining of wind noise affected signals). Also the output of a frequency bin selection (Y_{FBS}) approach, as stated in (6.20), is examined. The noise estimates, as derived in section 6.1, are used for the beamformer. Moreover, the ideal noise PSDs are used to get a benchmark. The short time recursive PSD smoothing was set to zero, since this achieved the best results due to the high non-stationarity of the wind noise.

Table 6.1: SSNR gain compared with single microphone for different beamformer outputs

Signal	Microphone Distance		
	7.1 mm	14.3 mm	21.4 mm
Y_{DS}	1.0 dB	1.2 dB	1.6 dB
Y_{FBS}	1.8 dB	2.3 dB	2.5 dB
Y_{MV} (noise estimate)	1.8 dB	2.3 dB	2.8 dB
Y_{MV} (noise benchmark)	1.8 dB	2.5 dB	3.0 dB

As can be observed, all beamformer approaches are able to improve the SSNR in the considered frequency region compared with a single microphone, where all SNR gains are getting larger as the distance between the microphones is increased. It is interesting to see that the delay-and-sum approach Y_{DS} has the worst performance for this scenario for all microphone distances. Note that Y_{DS} is the beamformer of the approach in [96] for this closely spaced microphone arrangement. The frequency bin selection approach shows quite similar results compared with the MV-beamformer that uses the noise estimates, which indicates that the short time PSDs at the microphones vary heavily. Comparing the performance using the estimated noise PSDs with that of the beamformer that uses the the actual noise PSDs, it is observed that the results regarding the SSNR are similar, i.e., the PSD estimates are sufficiently accurate.

6.2.3 Post Filter Output

Now the SSNR as well as the LSD for the complete MWF including the post filter are examined. To compare the post filter against other approaches, a wind noise reduction filter by Franz et. al. [88] that defines a filter function based on the magnitude squared coherence is used as a reference. The proposed post filter as well as the post filter derived in [88] are applied to the beamformer output Y_{MV}, which uses the noise estimates. As can be seen in Table 6.2, the SSNR can be further improved, while keeping the speech distortion below 1 dB compared with the single microphone signal Y_1.

Table 6.2: SSNR and LSD comparison for the post filter output

Signal	SSNR gain	LSD
Y_1	-	2.3 dB
Y_{MV}	2.8 dB	2.2 dB
Z	5.4 dB	3.3 dB
Y_{MV} + post filter after [88]	5.4 dB	3.3 dB

For the post filter comparison, the noise overestimation parameter was set to $\mu = 6$ and the short time PSDs used for the post filter, as well as the calculated MSC needed for the filter design in [88], were recursively smoothed by the same factor of 0.85 to make a fair comparison. As can be seen, both post filters are able to achieve the same amount of noise reduction with the LSD being in the same range for both approaches.

Figure 6.3 shows the spectrogram for the omnidirectional reference microphone as well as the output Z of the proposed wind noise reduction algorithm with a microphone distance of 21.4 mm. It can be observed that the high energetic noise terms in the low frequencies are successfully suppressed. Above 600 Hz the noise reduction is not as strong, i.e., the assumptions for the wind noise signal properties with this noise recording are only valid for frequencies below 600 Hz (cf. Fig. 6.2).

6.2.4 Wind Noise Only Scenario

Finally, the wind noise reduction is considered in a scenario containing only wind noise and no driving noise. Again, the beamformer output Y_{MV} with noise estimation is used with both post filter approaches as in section 6.2.3. All parameters except for the overestimation parameter are the same. The

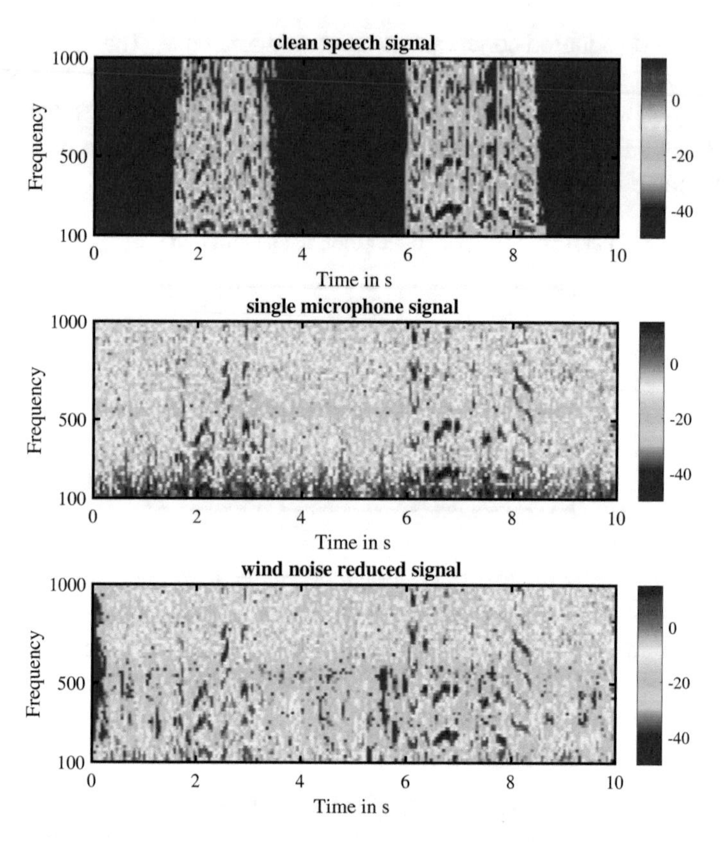

Figure 6.3: Spectrogram for the clean speech signal, the single microphone and the post filter output signal [14]

overestimation parameter for the Wiener post filter is $\mu = 8$. The SSNR of the single microphone Y_1 is 4.9 dB in this scenario. The results can be seen in Table 6.3.

A significant SSNR gain is achieved by the beamformer and the post filter outputs compared with the reference microphone. Comparing the results with the gains in Table 6.2, the achieved SSNR values are higher due to the absence of the driving noise. The deterioration of the speech distortion was kept below 1 dB compared with the single microphone signal Y_1 (LSD of 2.3 dB). This allows a fair comparison for both scenarios regarding the noise reduction of the post

Table 6.3: SSNR and LSD comparison for the algorithm output in a scenario containing only wind noise

Signal	SSNR gain	LSD
Y_1	-	2.3 dB
Y_{MV}	3.4 dB	2.3 dB
Z	9.3 dB	3.3 dB
Y_{MV} + post filter after [88]	9.2 dB	3.1 dB

filter. As can be observed, Z achieves a slightly higher SSNR gain than the combination of the MV beamformer and the post filter according to [88], but has a slightly higher LSD value.

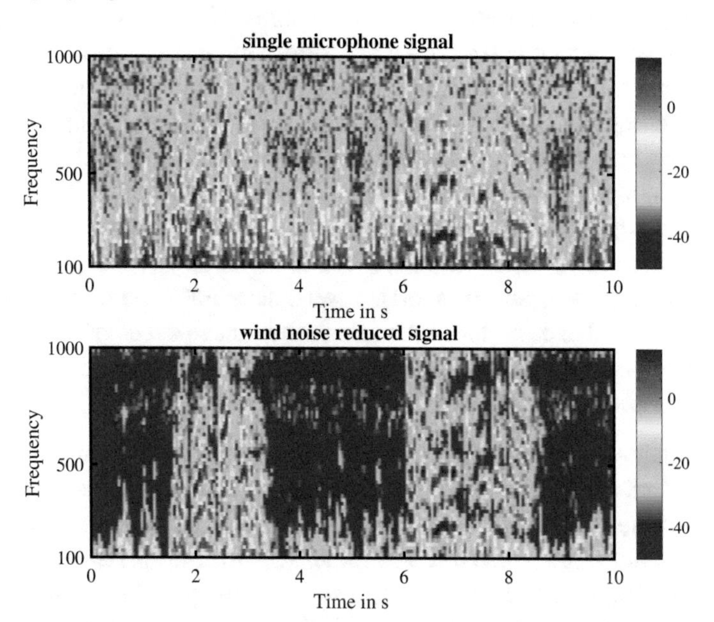

Figure 6.4: Spectrogram for a single microphone (top) and the post filter output signal (bottom) in a wind noise only scenario [14]

Figure 6.4 shows the spectrogram of the output Z for the wind noise only scenario. The noise is significantly reduced over a wide frequency range. Since the coherent driving noise terms are not present in this scenario, noise reduction can also be observed for frequencies above 600 Hz.

6.3 Summary

In this chapter, a wind noise reduction approach for a closely spaced microphone arrangement was proposed. Since differential beamforming is sensitive to wind noise, an alternative solution for noise reduction with closely spaced microphones was derived for situations where wind noise occurs. Although switching to one microphone in case of wind noise is mentioned in the literature as a possible solution, a multiple microphone setup is able to take advantage of the sound field diversity for even small microphone distances.

Exploiting the decomposition of the multichannel Wiener filter, the processing algorithm was implemented by a beamformer followed by a single channel Wiener post filter. The beamformer was derived based on the MVDR beamformer. For the derivation, the assumption about different signal properties of the speech and noise signals in the low frequencies for a closely spaced microphone array was used in combination with a known TDOA of the desired speech source. The speech signals were assumed to be identical for both microphones after the time alignment due to the large wave length in the low frequency range, whereas the short time wind noise PSDs were assumed to be decorrelated and unequal for each microphone. These assumptions led to the MV beamformer, which is basically a time and frequency dependent weighting of the microphone input signals.

The estimation of the speech and noise PSDs is commonly based on the assumption that the noise is stationary or varies only slowly, which is not the case for wind noise. Therefore, for the Wiener post filter as well as for the beamformer, the required speech and short time noise PSD estimates were derived based on a blocking of the speech source and a delay-and-sum beamformer. Both were used to obtain the final speech and noise PSD estimates.

To verify the proposed wind noise reduction algorithm and the assumptions about the signal properties, simulations were carried out using wind noise recordings of a closely spaced MEMS microphone arrangement in a car environment. It was shown that even for distances smaller than one centimeter the wind noise is decorrelated for frequencies below 600 Hz, while the speech signals are highly correlated. Regarding the SSNR, the output of the proposed MV beamformer is compared to the delay-and-sum beamformer and a selection approach that chooses a microphone channel for each frequency bin. As could be observed, all approaches were able to improve the SSNR compared with a single microphone, however, the delay-and-sum beamformer achieves the worst performance. This leads to the conclusion that the assumption about unequal and decorrelated short time noise PSDs holds. The SSNR gain comparison of the estimated PSDs with the ideal PSDs shows that the estimates are sufficiently good. The single channel Wiener post filter is able to improve the SSNR further and achieves similar results compared with state-of-the-art wind noise filters known in the literature.

Wind noise induced disruptions are a commonly known problem with differential beamforming. The differential beamforming references for the G-MWF, as proposed for noise reduction in hearing aids in the previous chapters, are therefore not suitable at all to remove wind noise artifacts. However, due to the closely spaced microphones, the proposed wind noise reduction approach may also be predestined for the application in hearing aids.

7 Background Noise Simulation based on MIMO Equalization

In the previous chapters, speech enhancement algorithms were developed that take the spatial information of the sound field into account by using more than one microphone signal. This raises the question how these algorithms can be tested regarding their enhancement capabilities. Additionally, often a comparison with state-of-the-art multiple microphone processing approaches that are already known in the literature is desired (i.e. [104, 105, 106]). For the simulations in previous chapters, the speech and noise signals were therefore recorded separately and mixed together afterwards. However, this simulations rely on the assumption that the microphones are perfectly linear and the speech and noise signals can be combined in terms of superposition.

For the evaluation of enhancement algorithms in noisy environments also background noise simulations are often used. In order to verify these algorithms in a real acoustic environment, the proper signal conditions between the recorded noise signals must be preserved. In the ETSI EG 202 396-1 standard [107], the background noise conditions are reproduced by several loudspeakers and one or two microphones. The acoustic transfer functions from the loudspeakers to the microphones are equalized in third octave bands and the time-difference-of-arrivals are compensated. However, for enhancement algorithms using multiple microphones this is not sufficient, since they take the spatial information of the noise field into account, which is not considered in this equalization approach.

In order to achieve a more accurate reproduction of the noise signals, the equalization must be capable to decorrelate the acoustic propagation paths of the simulation environment. This is considered in the ETSI TS 103 224 standard [108], where eight loudspeakers and microphones are used to recreate a sound field around a dummy head or a hand-held communication device. In the standard, the inverse of the acoustic transfer function matrix is used for

S. Grimm, *Directivity Based Multichannel Audio Signal Processing For Microphones in Noisy Acoustic Environments*, Schriftenreihe der Institute für Systemdynamik (IDS) und optische Systeme (ISO), https://doi.org/10.1007/978-3-658-25152-9_7

the equalization of the acoustic multiple input - multiple output (MIMO) system. While this approach shows good results compared with ETSI EG 202 396-1 and other approaches [109], several problems can occur if the inverse acoustic matrix is ill-conditioned. Therefore, a regularization factor is introduced, which can be optimized in different manners [110, 111].

In the following, an approach using more loudspeakers than microphones compared with [108] is proposed. Since it is possible to have access to the microphone signals for all noise reduction techniques derived in the previous chapters, the proposed approach aims to equalize the MIMO system at the actual microphones. In contrast, the ETSI TS 103 224 standard uses measurement microphones around the actual hand-held device or dummy head for the equalization.

By using a greater number of loudspeakers than microphones, multiple solutions for the inverse equalization matrix exist. One possible solution, the Moore-Penrose pseudo-inverse of the ATF matrix, can be calculated and used as a mixing matrix for the noise input signals. In the following, the impact of this pseudo-inverse on the equalization procedure is examined.

This chapter is outlined as follows. In section 7.1, the signal model and the notation of the MIMO acoustic environment is described as an expansion of the signal model in chapter 2.1. Further, the equalization approach that uses the pseudo-inverse of the acoustic transfer function matrix is derived in section 7.2, followed by the measurement results and the simulations which are presented in section 7.3.

This chapter has been published in [15].

7.1 The Signal Model

In this section, the acoustic MIMO model and its notation is presented. It is based on the signal model in chapter 2.1. However, the model needs to be expanded from the single input - multiple output (SIMO) system to the MIMO case, since more than one signal source (the number of required playback loudspeakers) is used. The MIMO system is considered as linear and time-invariant. It consists of N loudspeakers and M microphones ($N \geq M$). It is also assumed that the signal-to-noise ratio is sufficiently high, so noise influences can be neglected in the following. As a result, the signal $y_i(k)$ at the i-th microphone can be written as

$$y_i(k) = \sum_{n=1}^{N} h_{n,i}(k) * x_n(k)\,, \tag{7.1}$$

where $x_n(k)$ refers to the discrete input signal that is fed to the n-th loudspeaker and $h_{n,i}$ describes the acoustic impulse response from the n-th loudspeaker to the i-th microphone.

The acoustic transfer functions derived from the corresponding impulse responses as well as the MIMO input and output signals can be written in the frequency domain as the acoustic transfer function matrix

$$\mathbf{H}(\nu) = \begin{bmatrix} H_{11}(\nu) & H_{12}(\nu) & \cdots & H_{1M}(\nu) \\ H_{21}(\nu) & H_{22}(\nu) & \cdots & H_{2M}(\nu) \\ \vdots & \vdots & \ddots & \vdots \\ H_{N1}(\nu) & H_{N2}(\nu) & \cdots & H_{NM}(\nu) \end{bmatrix} \tag{7.2}$$

and furthermore the vectors

$$\mathbf{X}(\eta,\nu) = [X_1(\eta,\nu), X_2(\eta,\nu), \cdots, X_N(\eta,\nu)] \tag{7.3}$$
$$\mathbf{Y}(\eta,\nu) = [Y_1(\eta,\nu), Y_2(\eta,\nu), \cdots, Y_M(\eta,\nu)]\,, \tag{7.4}$$

where $\mathbf{X}(\eta,\nu)$ is the loudspeaker input signal vector and $\mathbf{Y}(\eta,\nu)$ is the microphone signal vector. In the following, the indices η and ν are omitted if possible.

As a result, the MIMO system equation, as given in (7.1), can be written in a compact form in the frequency domain

$$\mathbf{Y} = \mathbf{X}\mathbf{H}\,. \tag{7.5}$$

7.2 Equalization of the MIMO System

In the following, the equalization of the MIMO acoustic environment is derived. Therefore, a new signal vector $\mathbf{S}(\eta,\nu)$ is introduced, which contains the input signals (the recorded noise or speech reference signals) that need to be accurately reproduced at the microphones

$$\mathbf{S}(\eta,\nu) = [S_1(\eta,\nu), S_2(\eta,\nu), \cdots, S_M(\eta,\nu)]\,. \tag{7.6}$$

Furthermore, an equalization matrix $\mathbf{W}(\nu)$ is introduced that acts as a pre-equalization filter for the input signals before they are sent to the loudspeakers

$$\mathbf{W}(\nu) = \begin{bmatrix} W_{11}(\nu) & W_{12}(\nu) & \cdots & W_{1N}(\nu) \\ W_{21}(\nu) & W_{22}(\nu) & \cdots & W_{2N}(\nu) \\ \vdots & \vdots & \ddots & \vdots \\ W_{M1}(\nu) & W_{M2}(\nu) & \cdots & W_{MN}(\nu) \end{bmatrix}. \tag{7.7}$$

The whole system, from the input signals in vector $\mathbf{S}(\nu)$ to the microphone signals in vector $\mathbf{Y}(\nu)$, can be represented as a block diagram as shown in Figure 7.1.

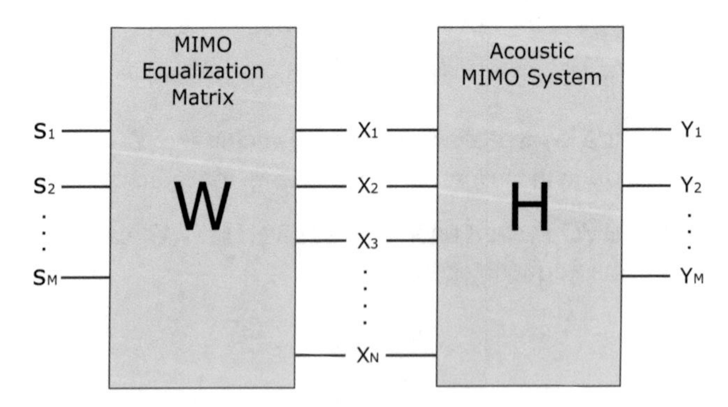

Figure 7.1: MIMO system - block diagram [15]

The relation between $\mathbf{S}(\nu)$ and $\mathbf{Y}(\nu)$ can be written as

$$\mathbf{Y} = \mathbf{SWH}\,. \quad (7.8)$$

In order to equalize the acoustic system, it must be assured that the following statement holds

$$\mathbf{Y} \stackrel{!}{=} \mathbf{S}\,. \quad (7.9)$$

This can be achieved by choosing $\mathbf{W}$ to fulfill

$$\mathbf{WH} = \mathbf{I}\,. \quad (7.10)$$

$\mathbf{I}$ denotes the unity matrix. If the number of loudspeakers is equal to the number of microphones ($N = M$), this obviously results in

$$\mathbf{W} = \mathbf{H}^{-1}\,, \quad (7.11)$$

if $\mathbf{H}$ has full rank and is well conditioned. In the ETSI TS 103 224 standard this is implemented as

$$\mathbf{W} = (\mathbf{H}^{\dagger}\mathbf{H} + \vartheta\mathbf{I})^{-1}\mathbf{H}^{\dagger}\,, \quad (7.12)$$

where † is the conjugate transpose and ϑ is a regularization factor to avoid problems due to an ill-conditioned matrix $\mathbf{H}$.

A matrix is ill-conditioned if it has a very high condition number κ, which is defined as the ratio between the maximal and minimal singular values σ_{max} and σ_{min}, respectively

$$\kappa(\nu) = \frac{\sigma_{max}(\nu)}{\sigma_{min}(\nu)}\,. \quad (7.13)$$

The condition number also relates the matrix norms of the matrices $\mathbf{H}$ and $\mathbf{W}$, i.e.,

$$\kappa(\nu) = ||\mathbf{H}|| \cdot ||\mathbf{W}||. \quad (7.14)$$

From (7.14) it can be observed that an ill-conditioned matrix $\mathbf{H}$ results in a matrix $\mathbf{W}$ with a large matrix norm. For large condition numbers, the regularization factor ϑ in (7.12) limits the norm of $\mathbf{W}$. However, the regularization prevents perfect equalization of the MIMO system.

For the proposed equalization approach, more loudspeakers than microphones are used ($N > M$). This reduces the probability of ill-conditioned matrices, as will be shown in the simulations in the following section. Hence, a smaller regularization factor ϑ can be chosen or the regularization is completely avoided.

With $N > M$ the matrices are not square and no unique inverse exists. By using the Moore-Penrose pseudo-inverse [112], an inverse of the acoustic system MIMO matrix $\mathbf{H}$ can be calculated, analogous to (7.12). This pseudo-inverse minimizes the matrix norm of the equalization matrix $\mathbf{W}$. The minimized norm also results in reduced power of the signals that are played back by the loudspeakers, which follows from the sub-multiplicative property

$$||\mathbf{X}|| \leq ||\mathbf{S}|| \cdot ||\mathbf{W}||. \tag{7.15}$$

7.3 Simulation and Measurement Results

In order to verify the proposed equalization approach to achieve proper reproduction of noise signals at the microphones, measurements in a car environment were taken. Therefore it is evaluated if the signal power as well as the spatial properties of the playback signals can be accurately reproduced at the microphones. Regarding the signal power evaluation, the PSDs at the microphones are compared with the PSDs of the input reference signals. To verify the accurate reproduction of the spatial properties, the magnitude squared coherence was chosen as a quality measure.

The simulation scenario consisting of four loudspeakers and two microphones is shown in Figure 7.2. The input signals are two noise signals that are recorded in a car at a driving speed of 100 km/h, which are aimed to be reproduced accurately at the microphones. For the equalization approach, the acoustic transfer functions in the simulation environment from each loudspeaker to each microphone were measured using a logarithmic sine sweep. Based on that measurement, the mixing matrix $\mathbf{W}$ was calculated by the Moore-Penrose pseudo-inverse. Since the elements in $\mathbf{W}$ contain non causal filters, they were transformed to the time domain and delayed to obtain a causal filter. Then the input signals were pre-equalized with the corresponding filters of the mixing matrix in the time domain. The filter length for the MIMO equalization was chosen to 8192 samples and a sampling frequency of $fs = 16$ kHz was used. The pre-equalized signals were sent to the loudspeakers and the results were recorded with the two microphones. For comparison, further measurements were also taken with the signals pre-equalized as suggested in the ETSI EG 202 396-1 standard.

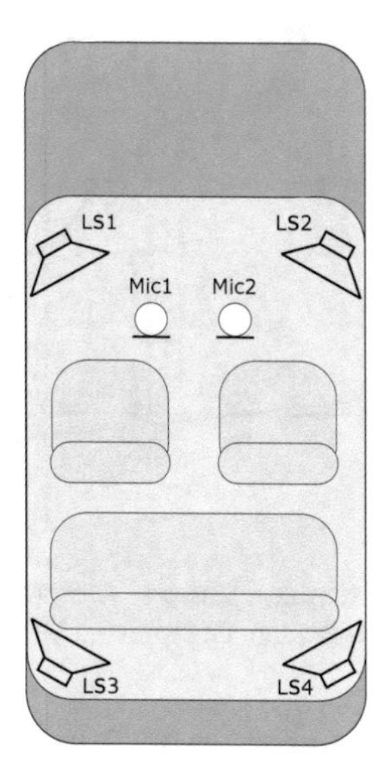

Figure 7.2: Loudspeaker-microphone arrangement in the car environment [15]

To verify if the correct power spectral densities of the input signals are reproduced at the microphones, the PSDs were analyzed in third octave bands for the proposed equalization approach as well as for the ETSI EG 202 396-1 standard and compared with the PSDs of the original input noise signals. The results are shown in Figure 7.3. The reference (A) shows the PSD for the input signal, which is aimed to be reproduced at microphone 1. The PSDs for the equalization after ETSI EG 202 396-1 (B) as well as for the proposed MIMO equalization approach (C) are shown. As can be seen, both equalization approaches match quite well in terms of the PSD compared with the reference signal.

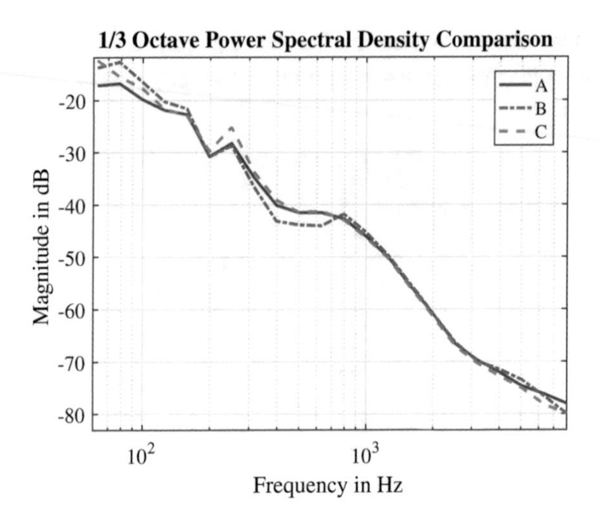

Figure 7.3: Power spectral density comparison - A: input signal (reference); B: ETSI EG 202 396-1 equalization; C: proposed MIMO equalization [15]

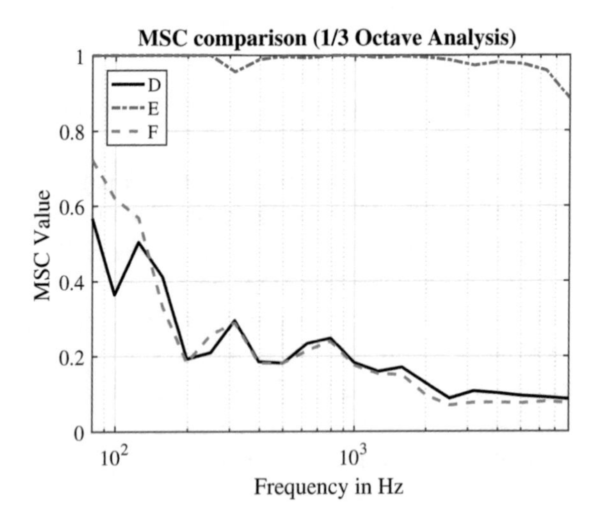

Figure 7.4: Magnitude squared coherence comparison - D: input signals (reference); E: ETSI EG 202 396-1 equalization; F: proposed MIMO equalization [15]

In Figure 7.4, a comparison for the correct reproduction of the magnitude squared coherence between the two input signals is depicted. The MSC for the input signals is denoted as the reference (D). In comparison, the MSC for the proposed MIMO equalization approach is shown (F). As can be observed, the MSC is reproduced quite well over the whole spectrum. In low frequencies, the performance decreases due to the not sufficient filter length of the equalization filters, but the overall performance can be considered to be good. The ETSI EG 202 396-1 standard (E) does not consider spatial reproduction at all, so the MSC is close to one for the observed frequencies.

To verify if the estimated acoustic transfer function matrices are well conditioned, the condition number over all frequencies for the matrices are plotted in Figure 7.5. The acoustic transfer function matrices are derived using two microphones and a varying number of loudspeakers ($M = 2$ and $N \in \{2,3,4\}$). The frequency dependent condition numbers for two (2×2), three (3×2) and all four loudspeakers (4×2) are shown. As can be observed, the frequency dependent condition number values are the highest for the two-loudspeaker-case and decrease as more loudspeakers are added. Hence, the measurements show that the inverse / pseudo-inverse pre-equalization matrix is better conditioned if more loudspeakers than microphones are used.

7.4 Summary

In this chapter, a background noise simulation approach was proposed to verify speech enhancement algorithms that use more than one microphone and take the spatial properties of the sound field into account. Therefore, a multiple-input multiple-output arrangement using more loudspeakers than microphones is considered as a simulation environment. In order to reproduce the input signals at the microphones correctly, the MIMO environment was equalized by the inverse acoustic transfer function matrix to decorrelate the acoustic signal paths. Since more loudspeakers than microphones are utilized, no unique inverse exists and the Moore-Penrose pseudo-inverse was applied since it minimizes the norm of the matrix. This also reduces the power required for the signal playback.

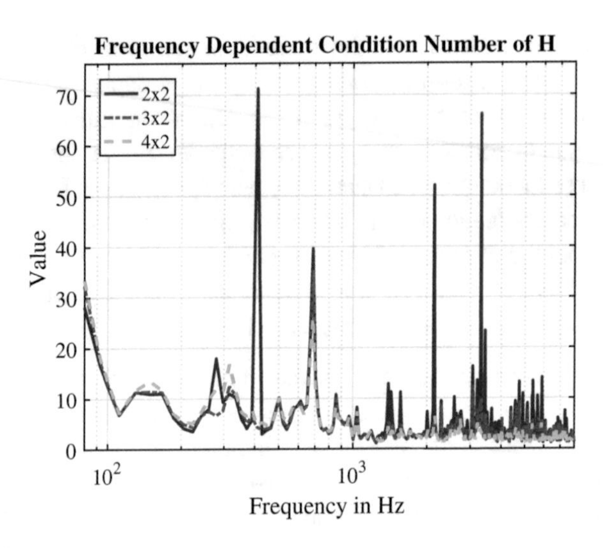

Figure 7.5: Condition number comparison - 2×2: two loudspeakers; 3×2: three loudspeakers, 4×2: four loudspeakers [15]

Measurements were taken in a car environment with up to four loudspeakers and two microphones to verify the proposed equalization approach. The PSDs of the reference noise input signals are compared with the PSDs of the microphone signals to show that the signal powers are correctly reproduced. In order to verify the accurate reproduction of the spatial properties between the microphones, the magnitude squared coherence function of the reference noise signals was compared with that of the microphone signals.

The measurements show that the reference noise PSDs are retained well at the microphones for the proposed equalization as well as for the ETSI EG 202 396-1 standard. For the MSC evaluation, the proposed background noise simulation approach results in similar MSC values compared with the input signals. However, slight deviations occur in the low frequencies due to the not long enough equalization filters. In comparison, the equalization using the ETSI EG 202 396-1 standard failed to reproduce the MSC completely. By varying the number of loudspeakers and therefore the dimension of the acoustic transfer function matrix, the influence on the condition number was examined. It was shown, that using more loudspeakers than microphones reduces the probability of an ill-conditioned matrix since the condition number decreases for the examined environment as more loudspeakers are added.

As a conclusion follows that the input signal properties can be reproduced sufficiently accurate at the microphones by the proposed background noise simulation approach. Using more loudspeakers than microphones also increases the probability of a well conditioned equalization matrix.

8 Conclusions

In this thesis, multichannel speech signal processing algorithms have been investigated. They all augment a desired speech signal, while reducing unwanted disruptions which are caused by background noise or reverberation. The usage of more than one microphone allows to take the spatial sound field into account to achieve a spatial filtering of the signals.

With the generalization of the multichannel Wiener filter, it was shown that the overall transfer function can be designed by a combining of the individual microphone channels. In contrast to previous research on the G-MWF, knowledge about the microphone arrangement as well as the location of a speaker are considered in this work. This allows to combine the narrow-band algorithms with broad-band beamforming approaches. As a result, a class of directivity based references for the generalized MWF has been derived in this thesis. It has been shown in simulations that these new references achieve an improvement regarding the broad-band SNR compared with the standard MWF for various application environments. For closely spaced microphones the differential beamforming references are able to improve the SNR by forming a directional response as the overall transfer function. For distributed microphones the proposed delay-and-sum based references are able to exploit the diversity of the sound field. Besides improved noise reduction compared with the standard MWF, these delay-and-sum based references reduce reverberation due to the coherent combining of the direct signal paths. In comparison with the approach that is based on a partial equalization of the magnitude of the acoustic transfer function [7], the proposed delay-and-sum based references show a similar performance regarding the noise reduction but with additional dereverberation capabilities.

However, the question remains which reference choice is appropriate for which microphone arrangement. In this work, equally spaced acoustic sensors have been investigated. Future research may address this by designing references

S. Grimm, *Directivity Based Multichannel Audio Signal Processing For Microphones in Noisy Acoustic Environments*, Schriftenreihe der Institute für Systemdynamik (IDS) und optische Systeme (ISO), https://doi.org/10.1007/978-3-658-25152-9

that consider signal combining for non-uniformly spaced arrays. This can potentially further improve the noise reduction and dereverberation.

Since the differential beamforming references are also applied in the context of hearing aids in the simulations, the preservation of the binaural cues is an important topic. The binaural multichannel Wiener filter is able to preserve the binaural cues of a desired speech signal perfectly. In this work, it has been examined if the reference designs that use a combining of the channels are still able to preserve the binaural cues of the desired speech source. Therefore the binaural MWF is generalized similar to the SDW-MWF. It was shown in simulations that the binaural cues can be preserved under the assumption that the acoustic transfer function ratios between the microphones of the left and the right hearing aid are pairwise identical. The standard binaural MWF can preserve the binaural cues of a desired signal source perfectly but distorts the binaural cues of the noise field. Existing noise cue preservation techniques can potentially be combined with the proposed directivtiy based references for the binaural Wiener filter in future research. This may achieve a superior noise reduction compared with the standard binaural MWF, while preserving the binaural cues of the noise field.

Differential beamforming and therefore also the differential beamforming references are sensitive to wind noise disruptions. This is caused by the correlation properties of the noise signals between adjacent microphones which are decorrelated for even small microphone distances. As a result, the wind noise terms are heavily amplified for differential beamforming processing approaches. In this thesis, a wind noise reduction algorithm for closely spaced microphones was derived, based on the decomposition of the MWF into a beamformer and a single channel post filter. Ordinary noise estimation procedures rely on the assumption that the noise is stationary or only varies slowly in time. This is not the case for wind noise which is highly non-stationary. Therefore an estimation approach was derived, which is based on the different signal properties of speech (highly correlated) and wind noise (decorrelated) between the microphones. The obtained estimates of the speech and noise PSDs are required for the beamformer as well as the single channel post filter. For the derivation of the beamformer, it was further assumed that the short time noise PSDs are unequal between the microphones for each frequency bin. Simulations with wind noise recordings of closely spaced MEMS microphones in a car show good results regarding the wind noise reduction. The proposed wind noise reduction algorithm for closely spaced microphones shows a superior performance compared with a recently introduced algorithm that assumes the noise PSDs to be identical at the microphones. The stationary driving noise, however, is not reduced by this approach since it is highly correlated for small microphone distances. Future research may address this by using the closely spaced microphones as a part of an array of more widely spaced microphones, where the spatial diversity of the sound field can be exploited for further noise reduction. Since the non-stationary noise terms are mostly reduced with the proposed approach, state-of-the-art noise estimation procedures can be chosen that rely on the assumption that the driving noise is only slowly varying.

Bibliography

[1] A. Spriet, M. Moonen, and J. Wouters, "Spatially pre-processed speech distortion weighted multi-channel Wiener filtering for noise reduction in hearing aids," in *International Workshop on Acoustic Echo and Noise Control (IWAENC)*, 2003, pp. 147–150.

[2] A. Spriet, M. Moonen, and J. Wouters, "Spatially pre-processed speech distortion weighted multi-channel Wiener filtering for noise reduction," *Signal Processing*, vol. 84, no. 12, pp. 2367–2387, 2004.

[3] S. Doclo, A. Spriet, M. Moonen, and J. Wouters, "Speech distortion weighted multichannel Wiener filtering techniques for noise reduction," in *Speech Enhancement*, chapter 9. Springer, Berlin, Heidelberg, New York, 2005.

[4] S. Doclo, A. Spriet, J. Wouters, and M. Moonen, "Frequency-domain criterion for the speech distortion weighted multichannel Wiener filter for robust noise reduction," *Speech Communication*, vol. 49, no. 7-8, pp. 636–656, July 2007.

[5] J. Chen, J. Benesty, Y. Huang, and S. Doclo, "New insights into the noise reduction Wiener filter," *IEEE Transactions on Audio, Speech and Language Processing*, vol. 14, no. 4, pp. 1218–1234, July 2006.

[6] T. C. Lawin-Ore, S. Stenzel, J. Freudenberger, and S. Doclo, "Generalized multichannel Wiener filter for spatially distributed microphones," in *Proc. ITG Conference on Speech Communication, Erlangen, Germany*, Sep 2014, pp. 1–4.

[7] S. Stenzel, T.C. Lawin-Ore, J. Freudenberger, and S. Doclo, "A multichannel Wiener filter with partial equalization for distributed microphones," in *IEEE Workshop on Applications of Signal Processing to Audio and Acoustics, Mohonk Mountain House, New Paltz, NY*, 2013.

S. Grimm, *Directivity Based Multichannel Audio Signal Processing For Microphones in Noisy Acoustic Environments*, Schriftenreihe der Institute für Systemdynamik (IDS) und optische Systeme (ISO)

[8] Darren B. Ward, Rodney A. Kennedy, and Robert C. Williamson, *Constant Directivity Beamforming*, pp. 3–17, Springer Berlin Heidelberg, Berlin, Heidelberg, 2001.

[9] Jacob Benesty and Chen Jingdong, *Study and Design of Differential Microphone Arrays*, Springer, Berlin, Heidelberg, 2013.

[10] S. Grimm, J. Freudenberger, T. C. Lawin-Ore, and S. Doclo, "Phase reference for the generalized multichannel Wiener filter," *EURASIP Journal on Advances in Signal Processing*, vol. 2016, no. 1, pp. 1 – 17, Jul 2016.

[11] S. Grimm and J. Freudenberger, "A phase reference for a multichannel Wiener filter by a delay and sum beamformer," in *Jahrestagung für Akustik (DAGA), Nürnberg*, Mar 2015, pp. 208–212.

[12] S. Grimm and J. Freudenberger, "A directivity based reference for the multichannel Wiener filter," in *Jahrestagung für Akustik (DAGA), München*, Mar 2018.

[13] S. Grimm, J. Freudenberger, and H. Schnepp, "Microphone diversity based wind noise reduction in a car environment using MEMS arrays," in *Jahrestagung für Akustik (DAGA), Kiel*, Mar 2017, pp. 1473–1476.

[14] S. Grimm and J. Freudenberger, "Wind noise reduction for a closely spaced microphone array in a car environment," *EURASIP Journal on Audio, Speech, and Music Processing*, vol. 2018, no. 1, pp. 1 – 7, Jul 2018.

[15] S. Grimm and J. Freudenberger, "Background noise simulation in cars based on multiple input – multiple output equalization," in *Jahrestagung für Akustik (DAGA), Kiel*, Mar 2017, pp. 299–302.

[16] Gary W. Elko, *Differential Microphone Arrays. In: Huang Y., Benesty J. (eds) Audio Signal Processing for Next-Generation Multimedia Communication System*, Springer, Boston, MA, 2004.

[17] Heinz Teutsch and Gary W. Elko, "First-and second-order adaptive differential microphone arrays," in *International Workshop on Acoustic Signal Enhancement*, 2001, pp. 35– 38.

[18] J. Benesty, M. Souden, and Y. Huang, "A perspective on differential microphone arrays in the context of noise reduction," *IEEE Transactions*

on *Audio, Speech, and Language Processing*, vol. 20, no. 2, pp. 699–704, Feb 2012.

[19] Gary W. Elko, "Microphone array systems for hands-free telecommunication," *Speech Communication*, vol. 20, no. 3, pp. 229 – 240, Dec 1996, Acoustic Echo Control and Speech Enhancement Techniques.

[20] J. Capon, "High-resolution frequency-wavenumber spectrum analysis," *Proceedings of the IEEE*, vol. 57, no. 8, pp. 1408 – 1418, Aug. 1969.

[21] E. A. P. Habets, J. Benesty, I. Cohen, S. Gannot, and J. Dmochowski, "New insights into the MVDR beamformer in room acoustics," *IEEE Transactions on Audio, Speech, and Language Processing*, vol. 18, no. 1, pp. 158–170, Jan 2010.

[22] E. A. P. Habets and J. Benesty, "A perspective on frequency-domain beamformers in room acoustics," *IEEE Transactions on Audio, Speech, and Language Processing*, vol. 20, no. 3, pp. 947–960, 2012.

[23] Joerg Bitzer and K. Uwe Simmer, *Superdirective Microphone Arrays*, pp. 19–38, Springer Berlin Heidelberg, Berlin, Heidelberg, 2001.

[24] S. Doclo and M. Moonen, "Superdirective beamforming robust against microphone mismatch," *IEEE Transactions on Audio, Speech, and Language Processing*, vol. 15, no. 2, pp. 617–631, Feb 2007.

[25] James W. Kates, *Digital Hearing Aids*, Plural Publishing, San Diego, CA, 2008.

[26] M. Er and A. Cantoni, "Derivative constraints for broad-band element space antenna array processors," *IEEE Transactions on Acoustics, Speech and Signal Processing*, vol. 31, no. 6, pp. 1378 – 1393, dec 1983.

[27] S. Gannot and I. Cohen, "Adaptive beamforming and postfiltering," in *Springer Handbook of Speech Processing*, J. Benesty, M. M. Sondhi, and Y. Huang, Eds., pp. 945–978. Springer Berlin Heidelberg, 2008.

[28] S. Wehr, I. Kozintsev, R. Lienhart, and W. Kellermann, "Synchronization of acoustic sensors for distributed ad-hoc audio networks and its use for blind source separation," in *Proceedings of IEEE Sixth International Symposium on Multimedia Software Engineering*, Dec 2004, pp. 18–25.

[29] S. Doclo, M. Moonen, T. Van den Bogaert, and J. Wouters, "Reduced-bandwidth and distributed MWF-based noise reduction algorithms for binaural hearing aids," *IEEE Transactions on Audio, Speech, and Language Processing*, vol. 17, no. 1, pp. 38–51, Jan 2009.

[30] T.C. Lawin-Ore and S. Doclo, "Analysis of rate constraints for MWF-based noise reduction in acoustic sensor networks," in *IEEE International Conference on Acoustics, Speech and Signal Processing (ICASSP)*, May 2011, pp. 269–272.

[31] S. Stenzel and J. Freudenberger, "Blind matched filtering for speech enhancement with distributed microphones," *Journal of Electrical and Computer Engineering*, Aug 2012, Article ID 169853, 15 pages.

[32] T. C. Lawin-Ore, S. Stenzel, J. Freudenberger, and S. Doclo, "Alternative formulation and robustness analysis of the multichannel Wiener filter for spatially distributed microphones," in *Proceedings of the International Workshop on Acoustic Signal Enhancement (IWAENC), Antibes, France*, Sep 2014, pp. 208–212.

[33] T. C. Lawin-Ore and S. Doclo, "Analysis of the average performance of the multichannel Wiener filter based noise reduction using statistical room acoustics," *Signal Processing*, vol. 107, pp. 96–108, Feb 2015.

[34] S. Markovich-Golan, A. Bertrand, M. Moonen, and S. Gannot, "Optimal distributed minimum-variance beamforming approaches for speech enhancement in wireless acoustic sensor networks," *Signal Processing*, vol. 107, pp. 4–20, Feb 2015.

[35] J. Schmalenstroeer, P. Jebramcik, and R. Haeb-Umbach, "A combined hardware-software approach for acoustic sensor network synchronization," *Signal Processing*, vol. 107, pp. 171–184, Feb 2015.

[36] S. Miyabe, N. Ono, and S. Makino, "Blind compensation of interchannel sampling frequency mismatch for ad hoc microphone array based on maximum likelihood estimation," *Signal Processing*, vol. 107, pp. 185–196, Feb 2015.

[37] L. Wang and S. Doclo, "Correlation maximization-based sampling rate offset estimation for distributed microphone arrays," *IEEE/ACM Trans.*

Audio, Speech and Language Processing, vol. 24, no. 3, pp. 571–582, Mar 2016.

[38] S. Gannot, D. Burshtein, and E. Weinstein, "Signal enhancement using beamforming and nonstationarity with applications to speech," *IEEE Transactions on Signal Processing*, vol. 49, no. 8, pp. 1614–1626, Aug 2001.

[39] T. C. Lawin-Ore and S. Doclo, "Reference microphone selection for MWF-based noise reduction using distributed microphone arrays," in *Proceedings of 10. ITG Symposium on Speech Communication*, Sept. 2012, pp. 31–34.

[40] Q.-G. Liu, B. Champagne, and P. Kaba, "Room speech dereverberation via minimum-phase and all-pass component processing of multi-microphone signals," *IEEE Pacific Rim Conference on Communications, Computers and signal Processing*, pp. 571–574, May 1995.

[41] J. B. Allen, D. A. Berlkey, and J. Blauert, "Multimicrophone signal-processing technique to remove room reverberation from speech signals," *The Journal of the Acoustical Society of America*, vol. 62, no. 4, pp. 912–915, Oct. 1977.

[42] E. A. P. Habets and J. Benesty, "A two-stage beamforming approach for noise reduction and dereverberation," *IEEE Transactions on Audio, Speech, and Language Processing*, vol. 21, no. 5, pp. 945–958, May 2013.

[43] I. Kodrasi and S. Doclo, "Joint dereverberation and noise reduction based on acoustic multichannel equalization," *IEEE Transactions on Audio, Speech, and Language Processing*, vol. 24, no. 4, pp. 680–963, April 2016.

[44] C. Knapp and G. Carter, "The generalized correlation method for estimation of time delay," *IEEE Transactions on Acoustics, Speech and Signal Processing*, vol. 24, no. 4, pp. 320–327, Aug 1976.

[45] G. C. Carter, *Coherence and Time Delay Estimation: An Applied Tutorial for Research, Development, Test and Evaluation Engineers*, IEEE Press, New Jersey, 1993.

[46] S. Doclo and M. Noonen, "Robust adaptive time delay estimation for speaker localization in noisy and reverberant acoustic environments," *EURASIP Journal on Applied Signal Processing*, pp. 1110–1124, Dec 2003.

[47] M. S. Brandstein and H. F. Silverman, "A robust method for speech signal time-delay estimation in reverberant rooms," in *Proceedings of IEEE International Conference on Acoustics, Speech and Signal Processing*, Apr 1997, pp. 375–378.

[48] T. G. Dvorkind and S. Gannot, "Time difference of arrival estimation of speech source in a noisy and reverberant environment," *Signal Processing*, vol. 85, no. 1, pp. 177–204, Jan 2005.

[49] J. Chen, J. Benesty, and Y. (Arden) Huang, "Performance of GCC- and AMDF-based time-delay estimation in practical reverberant environments," *EURASIP Journal on Applied Signal Processing*, vol. 2005, no. 1, pp. 25–36, Jan 2005.

[50] T. G. Manickam, R. J. Vaccaro, and D. W. Tufts, "A least-squares algorithm for multipath time-delay estimation," *IEEE Transactions on Signal Processing*, vol. 42, no. 11, pp. 3229–3233, Nov 1994.

[51] J. Chen, J. Benesty, and Y. Huang, "Time delay estimation in room acoustic environments: An overview," *EURASIP Journal on Applied Signal Processing*, vol. 2006, no. 1, pp. 1–19, May 2006.

[52] M Schwab, P Noll, and T Sikora, "Noise robust relative transfer function estimation," in *European Signal Processing Conference (EUSIPCO)*, Sept 2006, vol. 2, pp. 1–5.

[53] J. Sohn and W. Sung, "A voice activity detector employing soft decision based noise spectrum adaptation," in *IEEE International Conference on Acoustics, Speech and Signal Processing (ICASSP)*, May. 1998, pp. 365–368.

[54] J. Sohn, N. S. Kim, and W. Sung, "A statistical model-based voice activity detection," *IEEE Signal Processing Letters*, vol. 16, no. 1, pp. 1–3, Jan 1999.

[55] R. Martin, "Noise power spectral density estimation based on optimal smoothing and minimum statistics," *IEEE Transactions Speech and Audio Processing*, vol. 9, pp. 504–512, Jul. 2001.

[56] R. Martin, "Spectral subtraction based on minimum statistics," in *Proc. European Signal Processing Conference (EUSIPCO 1994)*, 1994, pp. 1182–1185.

[57] J. Freudenberger and S. Stenzel, "Time-frequency dependent voice activity detection based on a simple threshold test," in *IEEE Workshop on Statistical Sig. Proc. (SSP), Nice*, June 2011.

[58] Sumit Kumar and Rajat Gupta, "Estimating time delay using GCC for speech source localisation," *International Journal of Application or Innovation in Engineering and Management (IJAIEM)*, vol. 2, no. 5, pp. 24–30, May 2013.

[59] I. Céspedes, Y. Huang, J. Ophir, and S. Spratt, "Methods for estimation of subsample time delays of digitized echo signals," *Ultrasonic Imaging*, vol. 17, no. 2, pp. 142–171, 1995, PMID: 7571208.

[60] G. Jacovitti and G. Scarano, "Discrete time techniques for time delay estimation," *IEEE Transactions on Signal Processing*, vol. 41, no. 2, pp. 525–533, Feb 1993.

[61] I. Cohen, "Relative transfer function identification using speech signals," *Speech and Audio Processing, IEEE Transactions on*, vol. 12, no. 5, pp. 451–459, Sept 2004.

[62] S. Markovich-Golan, S. Gannot, and I. Cohen, "Multichannel eigenspace beamforming in a reverberant noisy environment with multiple interfering speech signals," *IEEE Transactions on Audio, Speech, and Language Processing*, vol. 17, no. 6, pp. 1071–1086, Aug 2009.

[63] S. Markovich-Golan and S. Gannot, "Performance analysis of the covariance subtraction method for relative transfer function estimation and comparison to the covariance whitening method," in *IEEE International Conference on Acoustics, Speech and Signal Processing (ICASSP)*, April 2015, pp. 544–548.

[64] R. Stewart and M. Sandler, “Database of omnidirectional and B-format room impulse responses,” in *IEEE International Conference on Acoustics Speech and Signal Processing (ICASSP)*, March 2010, pp. 165–168.

[65] M. R. Schroeder, “Frequency correlation functions of frequency responses in rooms,” *Journal of the Acoustical Society of America*, vol. 34, no. 2, pp. 1819–1823, Dec 1962.

[66] Patrick A. Naylor and Nikolay D. Gaubitch, *Speech Dereverberation*, Springer, London, 2010.

[67] J. H. L. Hansen and B. L. Pellom, “An effective quality evaluation protocol for speech enhancement algorithms,” in *International Conference on Speech and Language Processing*, 1998, pp. 2819–2822.

[68] H. Kayser, S. D. Ewert, J. Anemüller, T. Rohdenburg, V. Hohmann, and B. Kollmeier, “Database of multichannel in-ear and behind-the-ear head-related and binaural room impulse responses,” *EURASIP Journal on Advances in Signal Processing*, vol. 2009, no. 1, pp. 298605, Jul 2009.

[69] Sven Franz, Jörg Bitzer, and Uwe Simmer, “Multi-channel noise reduction for binaural hearing aids by using short-time spectral attenuation combined with noise estimators for non stationary noise,” in *International Conference on Acoustics, Rotterdam: including the 35th German Annual Conference on Acoustics (DAGA)*, Jun 2009, pp. 1–4.

[70] J. G. Desloge, W. M. Rabinowitz, and P. M. Zurek, “Microphone-array hearing aids with binaural output - Part 1: Fixed-processing systems,” *IEEE Transactions on Speech and Audio Processing*, vol. 5, no. 6, pp. 529–542, Nov 1997.

[71] J. Benesty, J. Chen, and Y. Huang, “Binaural noise reduction in the time domain with a stereo setup,” *IEEE Transactions on Audio, Speech, and Language Processing*, vol. 19, no. 8, pp. 2260–2272, Nov 2011.

[72] T. J. Klasen, S. Doclo, T. Van den Bogaert, M. Moonen, and J. Wouters, “Binaural multi-channel Wiener filtering for hearing aids: Preserving interaural time and level differences,” in *2006 IEEE International Conference on Acoustics Speech and Signal Processing Proceedings*, May 2006, vol. 5, pp. V–V.

[73] T. J. Klasen, T. Van den Bogaert, M. Moonen, and J. Wouters, "Binaural noise reduction algorithms for hearing aids that preserve interaural time delay cues," *IEEE Transactions on Signal Processing*, vol. 55, no. 4, pp. 1579–1585, April 2007.

[74] Simon Doclo, Thomas J. Klasen, Tim Van den Bogaert, Jan Wouters, and Marc Moonen, "Theoretical analysis of binaural cue preservation using multi-channel Wiener filtering and interaural transfer functions," in *Proceedings of the International Workshop on Acoustic Signal Enhancement (IWAENC), Paris, France*, Sept 2006, pp. 1–4.

[75] B. Cornelis, S. Doclo, T. Van dan Bogaert, M. Moonen, and J. Wouters, "Theoretical analysis of binaural multimicrophone noise reduction techniques," *IEEE Transactions on Audio, Speech, and Language Processing*, vol. 18, no. 2, pp. 342–355, Feb 2010.

[76] D. Marquardt, V. Hohmann, and S. Doclo, "Binaural cue preservation for hearing aids using multi-channel Wiener filter with instantaneous ITF preservation," in *2012 IEEE International Conference on Acoustics, Speech and Signal Processing (ICASSP)*, March 2012, pp. 21–24.

[77] D. Marquardt, V. Hohmann, and S. Doclo, "Coherence preservation in multi-channel Wiener filtering based noise reduction for binaural hearing aids," in *2013 IEEE International Conference on Acoustics, Speech and Signal Processing*, May 2013, pp. 8648–8652.

[78] D. Marquardt, V. Hohmann, and S. Doclo, "Interaural coherence preservation in multi-channel Wiener filtering-based noise reduction for binaural hearing aids," *IEEE/ACM Transactions on Audio, Speech, and Language Processing*, vol. 23, no. 12, pp. 2162–2176, Dec 2015.

[79] Daniel Marquardt, *Development and Evaluation of Psychoacoustically Motivated Binaural Noise Reduction and Cue Preservation Techniques*, Ph.D. thesis, Universität Oldenburg, 2015.

[80] D. Marquardt, E. Hadad, S. Gannot, and S. Doclo, "Theoretical analysis of linearly constrained multi-channel Wiener filtering algorithms for combined noise reduction and binaural cue preservation in binaural hearing aids," *IEEE/ACM Transactions on Audio, Speech, and Language Processing*, vol. 23, no. 12, pp. 2384–2397, Dec 2015.

[81] Jens Blauert, *Spatial Hearing: The Psychophysics of Human Sound Localization*, The MIT Press, Cambridge, Massachusetts, 2013.

[82] Frederic L. Wightman and Doris J. Kistler, "The dominant role of low-frequency interaural time differences in sound localisation," *The Journal of the Acoustical Society of America*, vol. 91, pp. 1648–61, 04 1992.

[83] E. Hadad, D. Marquardt, S. Doclo, and S. Gannot, "Binaural multichannel Wiener filter with directional interference rejection," in *2015 IEEE International Conference on Acoustics, Speech and Signal Processing (ICASSP)*, April 2015, pp. 644–648.

[84] Stuart Bradley, Tao Wu, Sabine von Hünerbein, and Juha Backman, "The mechanisms creating wind noise in microphones," in *Audio Engineering Society Convention 114*, Mar 2003.

[85] G. M. Corcos, "The structure of the turbulent pressure field in boundary-layer flows," *Journal of Fluid Mechanics*, vol. 18, no. 3, pp. 353–378, 1964.

[86] Christoph Matthias Nelke and Peter Vary, "Dual microphone wind noise reduction by exploiting the complex coherence," in *Proceedings of Speech Communications - 11. ITG Symposium*, Sep 2014.

[87] D. Keith Wilson and Michael J. White, "Discrimination of wind noise and sound waves by their contrasting spatial and temporal properties," *Acta Acustica United with Acustica*, vol. 96, no. 96, pp. 991–1002, Nov 2010.

[88] Sven Franz and Joerg Blitzer, "Multi-channel algorithms for wind noise reduction and signal compensation in binaural hearing aids," in *International Workshop on Acoustic Signal Enhancement (IWAENC)*, Aug 2010.

[89] Christoph Matthias Nelke and Peter Vary, "Measurement, analysis and simulation of wind noise signals for mobile communication devices," in *International Workshop on Acoustic Signal Enhancement (IWAENC)*, Sep 2014.

[90] Christoph Matthias Nelke, Navin Chatlani, Christophe Beaugeant, and Peter Vary, "Single microphone wind noise PSD estimation using signal centroids," in *IEEE International Conference on Acoustic, Speech and Signal Processing (ICASSP)*, May 2014.

[91] Shingo Kuroiwa, Youji Mori, Satoru Tsuge, Masashi Takashina, and Fuji Ren, "Wind noise reduction method for speech recording using multiple noise templates and observed spectrum fine structure," in *International Conference on Communication Technology*, Nov 2006.

[92] Brian King and Les Atlas, "Coherent modulation comb filtering for enhancing speech in wind noise," in *Proceedings of International Workshop on Acoustic Signal Enhancement (IWAENC)*, Sep 2008.

[93] Elias Nemer and Wilf Leblanc, "Single-microphone wind noise reduction by adaptive postfiltering," in *Proceedings of IEEE Workshop on Applications of Signal Processing to Audio and Acoustics (WASPAA)*, Oct 2009.

[94] Christian Hofman, Tobias Wolff, Markus Buck, Tim Haulik, and Walter Kellermann, "A morphological approach to single-channel wind-noise suppression," in *Proceedings of International Workshop on Acoustic Signal Enhancement (IWAENC)*, Sep 2012.

[95] Christoph Matthias Nelke, Niklas Nawroth, Marco Jeub, Christophe Beaugeant, and Peter Vary, "Single microphone wind noise reduction using techniques of artificial bandwidth extension," in *Proceedings of European Signal Processing Conference (EUSIPCO)*, Aug 2012.

[96] Philipp Thüne and Gerald Enzner, "Maximum-likelihood approach to adaptive multichannel-Wiener postfiltering for wind-noise reduction," in *ITG Conference on Speech Communication*, Oct 2016.

[97] M. Turqueti, J. Saniie, and E. Oruklu, "MEMS acoustic array embedded in an FPGA based data acquisition and signal processing system," in *2010 53rd IEEE International Midwest Symposium on Circuits and Systems*, Aug 2010, pp. 1161–1164.

[98] Ines Hafizovic, Carl-Inge Colombo Nilsen, Morgan Kjølerbakken, and Vibeke Jahr, "Design and implementation of a mems microphone array system for real-time speech acquisition," *Applied Acoustics*, vol. 73, no. 2, pp. 132 – 143, 2012.

[99] Jelmer Tiete, Federico Domínguez, Bruno da Silva, Laurent Segers, Kris Steenhaut, and Abdellah Touhafi, "Soundcompass: A distributed mems microphone array-based sensor for sound source localization," *Sensors*, vol. 14, no. 2, pp. 1918–1949, 2014.

[100] Gary Elko, "Small directional microelectromechanical systems (mems) microphone arrays," *Proceedings of Meetings on Acoustics*, vol. 19, no. 1, pp. 030033, 2013.

[101] A. Palla, L. Fanucci, R. Sannino, and M. Settin, "Wearable speech enhancement system based on MEMS microphone array for disabled people," in *2015 10th International Conference on Design Technology of Integrated Systems in Nanoscale Era (DTIS)*, April 2015, pp. 1–5.

[102] J. Freudenberger, S. Stenzel, and B. Venditti, "Spectral combining for microphone diversity systems," in *Proc. European Signal Processing Conference (EUSIPCO), Glasgow*, Aug 2009, pp. 854–858.

[103] R. Zelinski, "A microphone array with adaptive post-filtering for noise reduction in reverberant rooms," in *ICASSP-88, International Conference on Acoustics, Speech, and Signal Processing*, Apr 1988, pp. 2578–2581 vol.5.

[104] P. Vary and R. Martin, *Digital Speech Transmission: Enhancement, Coding and Error Concealment*, Wiley & Sons, Chichester, UK, 2006.

[105] Jacob Benesty, Jingdong Chen, and Yiteng Huang, Eds., *Microphone Array Signal Processing*, Springer Berlin Heidelberg, 2008.

[106] Yiteng Huang, Jacob Benesty, and Jingdong Chen, Eds., *Acoustic MIMO Signal Processing*, Springer Berlin Heidelberg, 2006.

[107] European Telecommunication Standards Institute, "Speech processing, transmission and quality aspects - part 1: Background noise simulation technique and background noise database, vol. 4," *ETSI EG 202 396-1*, pp. 1–58, 2011.

[108] European Telecommunication Standards Institute, "A sound field reproduction method for terminal testing including background database, vol. 1," *ETSI TS 103 224*, pp. 1–36, 2014.

[109] Juan David Gil Corrales, Marton Marschall, Torsten Dau, Wookeun Song, Claus Blaabjerg, Michael Hoby Andersen, and Soren W. Christensen, "Simulation of realistic background noise using multiple loudspeakers," in *Danish Sound Innovation Network*, 2014.

[110] Scott G. Norcross and Martin Bouchard, "Multichannel inverse filtering with minimal-phase regularization," in *Audio Engineering Society Convention 123, 1-8*, Oct 2007.

[111] Juan David Gil Corrales, Wookeun Song, and Ewen Macdonald, "Reproduction of realistic background noise for testing telecommunications devices," in *Audio Engineering Society Convention 182, 1-10*, May 2015.

[112] Medhat A. Rakha, "On the Moore–Penrose generalized inverse matrix," *Applied Mathematics and Computation*, vol. 158, no. 1, pp. 185 – 200, 2004.

Printed in the United States
By Bookmasters